The New Math

The New Math

A Political History

CHRISTOPHER J. PHILLIPS

THE UNIVERSITY OF CHICAGO PRESS   CHICAGO AND LONDON

CHRISTOPHER J. PHILLIPS is currently assistant professor and faculty fellow in New York University's Gallatin School of Individualized Study and has been appointed assistant professor in Carnegie Mellon University's Department of History.

The University of Chicago Press, Chicago 60637
The University of Chicago Press, Ltd., London
© 2015 by The University of Chicago
All rights reserved. Published 2015.
Printed in the United States of America
24 23 22 21 20 19 18 17 16 15   1 2 3 4 5

ISBN-13: 978-0-226-18496-8 (cloth)
ISBN-13: 978-0-226-18501-9 (e-book)
DOI: 10.7208/chicago/ 9780226185019.001.0001

Library of Congress Cataloging-in-Publication Data

Phillips, Christopher J. (Christopher James), 1982– author.
  The new math : a political history / Christopher J. Phillips.
    pages ; cm
  Includes bibliographical references and index.
  ISBN 978-0-226-18496-8 (cloth : alk. paper) — ISBN 978-0-226-18501-9 (e-book)
1. Mathematics—Study and teaching—Political aspects—United States.
2. Mathematics—Study and teaching—United States—History—20th century.
3. Education and state—United States—History—20th century.   I. Title.
  QA13.P49 2015
  510.71'073—dc23
                                                            2014012587

♾ This paper meets the requirements of ANSI/NISO Z39.48–1992 (Permanence of Paper).

FOR MY FAMILY

Contents

CHAPTER 1. Introduction: The American Subject   1

CHAPTER 2. The Subject and the State: The Origins of the
New Math   22

CHAPTER 3. The Textbook Subject: Mathematicians and the
New Math   47

CHAPTER 4. The Subject in Itself: Arithmetic as Knowledge   75

CHAPTER 5. The Subject in the Classroom: The Selling of the
New Math   96

CHAPTER 6. The Basic Subject: New Math and Its Discontents   121

Epilogue   145

Acknowledgments   151

Notes   153

Bibliography   199

Index   231

CHAPTER ONE

Introduction

The American Subject

Hooray for new math,
New-hoo-hoo-math,
It won't do you a bit of good to review math.
It's so simple,
So very simple,
That only a child can do it!

S atirist Tom Lehrer's song "New Math" captured for many Ameri-cans the absurdity and complexity of midcentury reforms in math classrooms. The tune was included on his 1965 album *That Was the Year That Was*, the cover of which prominently featured contemporary news-paper headlines. Lehrer's track listing stands as a record of the era's cul-tural history: "Send the Marines," "Wernher von Braun," "Vatican Rag." Lehrer's "New Math" spoof purported to be a lesson for parents con-fused by recent changes in their children's arithmetic textbook, pointing out that success in the new curriculum no longer required getting the an-swer right, only "understand[ing] what you are doing." Lehrer informed listeners that some new math problems involved base eight instead of the usual base ten (or decimal) system. Luckily, base eight was "just like base ten really—*if you're missing two fingers.*"

"New Math" was both a joke and a comment on the products of a controversial curriculum project, partially sponsored by the National Science Foundation (NSF) and involving hundreds of mathematicians, teachers, education professors, and administrators. Deemphasizing rote calculation while infamously introducing sets and other new concepts, the designers of the new math attempted to fundamentally reform the way Americans thought about mathematics. The curriculum rose and fell swiftly: initially introduced into schools between 1958 and 1962, the new math's influence peaked in the 1965 school year and was widely condemned a decade later. Newspapers, television shows, and comic strips all took notice of the reforms. Parents pored through algebra and arithmetic textbooks for the first time in years, and taxpayers hotly debated the multi-million-dollar reforms that they had, indirectly, paid for. Lehrer, himself a onetime teacher of mathematics, accurately included the new math among the major events of midcentury America.

There was, however, no such stable and coherent thing as *the* new math. The label loosely refers to a collection of curriculum projects, throughout the 1950s and 1960s, whose approaches—and resulting textbooks—diverged substantially, both mathematically and pedagogically. The phrase "new math" emerged around 1960 as shorthand for these "new mathematics curricula." A range of awkward acronyms for math reform groups—from UICSM and MINNEMAST to GCMP and SMSG—appeared in this period, and one perhaps overly optimistic estimate suggested that by 1965 at least half the nation's students were using new math textbooks.[1] At a minimum, millions of new books entered midcentury math classrooms.

Nearly all the reformers were driven by a sense that the way math had been taught for decades was no longer working. In an age of increasingly sophisticated mathematical models and technological systems—and just after the mathematical sciences had contributed so spectacularly to the allied victory in the Second World War—the idea of math as a set of facts and esoteric techniques appeared outmoded. Moreover, reformers believed math class provided intellectual training. Existing textbooks were not factually incorrect or ineffective per se but were cultivating the "wrong" mental habits in students. Teaching mathematics as rote memorization of multiplication tables meant teaching students the wrong way to think about mathematics and, more importantly, the wrong way to reason in general.

One reform program was by far the most influential in the period: the School Mathematics Study Group (SMSG). As the primary recipient of federal funding and as an initiative founded by joint action of the professional organizations of mathematicians and mathematics teachers, SMSG effectively created the "official" version of new math. Led by mathematician Edward Begle, the group worked from 1958 to 1972 to produce textbooks, monographs, teacher training guides, and a variety of other educational materials. With the imprimatur of the federal government and of professional mathematicians, Begle was ever mindful of the need to avoid appearing as if SMSG were imposing a curriculum on schools. As a result, SMSG produced only temporary and disposable materials, ones that were intended to model reforms for teachers and publishers without directly challenging the sale of commercial textbooks. Given the size and complexity of the student population studying mathematics at any one time, SMSG's leaders thought that it would be better to influence publishers with exemplary books than to create competing textbooks.[2] The actual effect of SMSG was nevertheless substantial, both in the extent to which its own books were used and in the direct influence its participants had on the construction of commercial textbooks.

Although originally funded to work on textbooks for the "college capable" students in secondary schools, SMSG gradually expanded its operation, producing textbooks for every grade and type of student, including material for elementary schools, "culturally disadvantaged"—mainly inner-city—children, and "slow" students. SMSG alone published nearly four million copies of over twenty-six different textbooks, in addition to teachers' manuals and monographs.[3] By the late 1960s, SMSG was the dominant organization in mathematics curriculum design, serving as a clearinghouse for various curricular reforms and education initiatives, as well as providing the infrastructure by which new mathematics curricula were tested.

SMSG's initial goal of reaching the talented students who had failed to be attracted to the field was a concern of great importance in an era when scientific innovation and military success were understood to go hand-in-hand. SMSG's leaders never thought, however, that their efforts would substantially increase the number of American mathematicians. As Begle explained, "The number of high school students, even if we consider only the better twenty-five percent, who go on to become research mathematicians is so infinitesimal that we spend almost no time worrying about them." Rather, "we must give as many students as possible a solid foundation in mathematics so that they will not be handicapped in later years, no matter what occupation they choose."[4] Given that "no one can predict what mathematical skills will be important and useful in the future," Begle wanted SMSG to ensure that *all* students possessed "an understanding of the role of mathematics in our society" because it was essential for "intelligent citizenship."[5]

The idea that math should be taught as a central component of intelligent citizenship was integral to the rise and fall of the new math. Many of the new textbooks, regardless of whether they emerged directly from SMSG or from one of the other new math groups, promoted mathematics' role in liberal education. The new math's development and deployment were based upon claims about the special nature of "modern" mathematical knowledge, the relationship between this nature and the mental habits resulting from its study, and the importance of these particular habits for the shaping of U.S. citizens. Math class was said to provide epistemological training—teaching students about what counts as valid knowledge and the grounds for its validity. In turn, the new math was rejected in the 1970s primarily because the arguments put forth in the late 1950s and 1960s about the ability of "modern" mathematics to promote intellectual discipline ceased to be compelling. The backlash certainly didn't entail rejection of the idea that learning mathematics counted as learning to think. Critics of the new math simply put forward rival arguments about the relationship between mathematical knowledge and intellectual habits. Supporters and critics of the new math had different conceptions of the merits and qualities of particular ways of thinking, that is, different conceptions of the mental discipline virtuous citizens should possess.

*   *   *

This book is a political history of the new math, one that grounds and interrogates midcentury American history through the changing mathematics curriculum.[6] It is a political history, but not because it is mainly concerned with connecting the curriculum to specific political platforms. Rather, proponents and opponents of the new math believed the curriculum could order and shape the mind, the family, the society, and the state. The reform of the mathematics curriculum was never limited to discussions about which topics to cover; the curriculum always entailed an argument about the proper relationship between the content and purpose of education. Mathematicians at midcentury did not agree about the nature of their subject, and Americans certainly disagreed about the mental habits math class ought to promote. Furthermore, the intellectual claims made about the curriculum and the role of the schools changed dramatically between 1955 and 1975. As a result, Americans evaluated the math curriculum differently. The new math embedded, instantiated, and made visible the changing politics of midcentury America.

Questions of why the curriculum succeeded or failed are important and have engaged subsequent curriculum designers and critics. The focus of *The New Math* is instead on grounding the perception of success and failure in changing evaluations of the nature of mathematical knowledge, and of the relevance of particular habits of thought for the cultivation of virtuous citizens. This approach calls into question one typical story of the new math's demise: that its supporters failed to deliver on the promise of improving computation skills. Neither the curriculum's designers nor its supporters ever focused on whether the new math would improve calculation ability. They talked instead about needing to prepare citizens for modern society, for a world of complex challenges, seemingly rapid technological changes, and unforeseeable future conflicts. Critics' gestures toward declining computation ability not only relied upon flimsy evidence but also overlooked the fact that the curriculum was never intended to improve the percentage of students who knew multiplication tables by heart. What had changed was a complicated set of *political* commitments, concerning the value of mechanistic intellectual habits, the relative importance of elite forms of knowledge compared to local and traditional ones, and the role of mathematics as mental discipline—that is, commitments concerning the way learning mathematics counted as learning to think.

A Discipline That Disciplines

Many fields of knowledge are useful, conveying practical information or manipulative skills. Mathematics is among these subjects, providing a collection of techniques used widely to measure and model the world. Its methods, nomenclature, and authority are claimed by a wide assortment of expert practices, from wine assessment and election polling to meteorology. Math provides the language for the quantification of certainty and the concepts that give structure to scientific descriptions of the universe. Yet, math is also abstract, celebrated for its insularity from the mundane and messy facts of the physical world. It is in this sense a chameleon of a discipline. It is obvious and evident—what else could $2 + 2$ be but 4?—as well as abstract and obtuse—what could it possibly mean to claim that $e^{i\pi} = -1$? Mastery requires years of study within esoteric domains, but most baseball fans are familiar with how to calculate a batting average. Mathematical practices are ubiquitous and obscure, quotidian and esoteric.

If many fields are useful, only a very few subjects are said to be good for the mind, preparing the student to think well in dealing with a wide variety of life's intellectual, social, moral, and political predicaments. Among these subjects, mathematics has long enjoyed a special place. Math class is meant to convey information about triangles, numbers, and equations, but also to provide mental exercise; it has never been just about learning facts. Math is a discipline which disciplines.

Math has, after all, been associated with pure reason and reliable knowledge at least since Greek antiquity. The term *mathematics* had a much more expansive meaning in its original Greek context, referring to general disciplines or subjects of study. For centuries, mathematics and the related term *mathesis* encompassed both elite practices of deductive or symbolic reckoning and the useful skills of surveying, celestial navigation, astrology, and harmonics.[7] Even as its methods, objects, and uses have changed, mathematics has remained a collection of practical techniques known by many as well as a compilation of esoteric results understood by only a few.

Ancient Greek geometry was the most successful practice holding a special claim on reasoning. Deductive geometry, as historian Reviel Netz has suggested, with its diagrams, highly specific language, and strict conventions, was effectively an "idealized, written version of oral argument." Netz's *The Shaping of Deduction in Greek Mathematics* concludes that mathematicians were likely "eccentrics" in a "world of doctors, sophists, and rhetoricians"—but eccentrics who established stable practices for making formal rhetorical arguments.[8] Greek geometry was a practice impossible to situate historically without an understanding of the contemporary intellectual and material tools of persuasive reasoning.

Take, for example, one of the most well-known ancient instances of math standing proxy for reasoning. In Plato's *Meno*, Socrates shows his interlocutor how even an uneducated slave "already knows" how to construct a square that has double the area of a given square. He leads the slave to build the square using the diagonal of the original square, then continues:

SOCRATES: Has he answered with any opinions that were not his own?
MENO: No, they were all his.

SOCRATES: Yet he did not know, as we agreed a few minutes ago.

MENO: True.

SOCRATES: But these opinions were somewhere in him, were they not?

MENO: Yes.

SOCRATES: So a man who does not know has in himself true opinions on a subject without having knowledge.

MENO: It would appear so.

SOCRATES: At present these opinions, being newly aroused, have a dreamlike quality. But if the same questions are put to him on many occasions and in different ways, you can see that in the end he will have a knowledge on the subject as accurate as anybody's.

MENO: Probably.

SOCRATES: This knowledge will not come from teaching but from questioning. He will recover it for himself.[9]

Geometry was not the subject of particular facts so much as the exemplar of intuition.

In the early modern period, as in antiquity, math was still tied to intellectual and philosophical practices, even if very few individuals ever learned more than simple arithmetic. Historian Matthew Jones has shown how René Descartes, Blaise Pascal, and Gottfried Leibniz all conceived mathematics as a way of cultivating the self. Mathematical techniques improved the ability of humans to reason and examine evidence, they claimed. Near contemporaries opposed extensive training in or reliance on mathematics with precisely the opposite reasoning—math, they suggested, was unsuited for scientific inquiry or individual development.[10] Mathematics' contingent relationship to intellectual training grounded the arguments for and against its use.

The association between math and reasoning was not exclusively dependent upon the special nature of geometric knowledge. Probability, algebra, and analysis were all developed in part on the presumption that the results of mathematical demonstrations and human intuition ought to align. This was said to be true as both a principle of contemporary practice and a fact borne out historically. The eighteenth-century mathematician Jean-Étienne Montucla declared that a well-done history of mathematics "could be looked upon as a history of the human mind, since it is in this science more than all others that man makes known the excellence of the gift of intelligence which God has given him to raise him above all other creatures." Montucla's sentiment drew on a long tradition emphasizing that God wrote the "book of nature" in the language of mathematics, and then uniquely bestowed upon humans the ability to learn and read this language.[11] Doing mathematics meant developing the divine gift of reason.

Debates about what mathematics to teach in nineteenth-century Britain were likewise adjudicated on the basis of how mathematics was understood to train students' minds. These resolutions—with their accompanying pedagogical transformations—in turn determined the emphasis of academic research.[12] The appointment of the University of Edinburgh chair of mathematics in the 1830s made the stakes clear. The candidates, Philip Kelland and Duncan Gregory, espoused two different mathematical practices. Kelland preferred algebraic or symbolic methods and Gregory emphasized geometric ones. The controversy over the nomination extended beyond the usual academic wrangling, because the mathematical distinctions were believed to map onto differences in the kind of mental habits cultivated in future students. One of the most influential Scottish philosophers of the period, Sir William Hamilton, wrote in support of Gregory:

> The mathematical process in the symbolical method [i.e., the algebraic] is like running a rail-road through a tunnelled mountain; that in the ostensive [i.e., the geometrical] like crossing the mountain on foot. The former carries us, by a short and easy transit, to our destined point, but in miasma, darkness and torpidity, whereas the latter allows us to reach it only after time and trouble, but feasting us at each turn with glances of the earth and of the heavens, while we inhale health in the pleasant breeze, and gather new strength at every effort we put forth.[13]

Hamilton saw differences in mathematical method as differences in the speed and quality of reasoning. The relative worth of mathematical practices was based on the mental habits inculcated, not on the conclusions reached.

The notion of mathematics as a mode of revealing and developing reason persists. Many institutions rely upon mathematics exams to rank individuals for intellectual honors. Standardized and "high stakes" examinations remain central meritocratic mechanisms for evaluating teachers, reducing inequality, and identifying underutilized talent. The American SAT requires knowledge of geometry, arithmetic, and algebra, but nothing of history, poetry, or chemistry.[14] In a 2013 *Atlantic* article promoting the latest mathematics standards—which form part of the Common Core standards—the author claimed math class was a place for developing reasoning skills, not for conveying facts: "In our new technological world, employers do not need people who can calculate correctly or fast, they need people who can reason about approaches, estimate and verify results, produce and interpret different powerful representations, and connect with other people's mathematical ideas." A successful math curriculum was one that promoted successful reasoning ability generally.[15]

Pedagogy and Habitus

Math must still be figured out. No matter how much teachers insist that addition or differentiation is logical, students continue to get the wrong answer even after seeing countless exemplars, rules, and patterns. It simply isn't obvious how one "goes on." Math is intuitive only to the extent that one has the *right* intuition; it is empirical only if students learn *how* to count. The corrective marks on the page or board, the tone used to ask questions or to reject suggested solutions, the praise given or withheld, the diagrams and rules—the order of the subject is meant to order the thinking and behavior of the students.

It is not the case that certain forms of classroom organization, pedagogical techniques, or textbook content effectively "discipline" while others do not. Every classroom involves a "hidden curriculum," where students learn the structure and practices of the institution alongside the content.[16] The math curriculum may be intended to train students to think rigorously, loosely, creatively, flexibly, linearly, or any way at all, so long as it happens to be valued. The school may be designed to produce individuals who internalize the militaristic order of the hourly bells and the rigid hierarchical structure of the classroom—or ones who are self-motivated and believe learning to be fun and playful and experimental.

Different intellectual habits are assumed to define different sorts of people. They can serve, in philosopher Michel Foucault's terminology, as a technique or technology of the self. These were "techniques which permit individuals to effect, by their own means, a certain number of operations on their own bodies, on their own souls, on their own thoughts, on their own conduct, and this in a manner so as to transform themselves, modify themselves."[17] One evocative example of the possible effects of intellectual habits was detailed by art historian Erwin Panof-

sky in his 1948 assessment of Gothic architecture. Panofsky surmised
that a "genuine cause-and-effect relationship" existed between scholasticism and the architecture of Gothic cathedrals—that the mental habits
learned in schools eventually gave rise to the specific design of twelfth-
and thirteenth-century cathedrals. Even with the limitations and pre-
dictable counterexamples to such an argument, sociologist Pierre Bourdieu (among others) would appropriate similar ideas in defining *habitus*
as a way of explaining the relationship between individual subjectivity
and social structures.[18] Educational institutions—and the curriculum in
them—shape social order through the cultivation of individual habits
and dispositions.

An increasing number of scholars over the last decades have similarly
emphasized the importance of the tools, materials, practices, and sites
of intellectual training. Following the work of Thomas Kuhn, historians
such as David Kaiser and Andrew Warwick have argued for moving beyond conceiving classrooms in purely negative terms—thinking of techniques of the self as constricting or punitive—to consider the *productive*
ways pedagogy cultivates material and intellectual practices.[19]

The importance of the role intellectual habits were assumed to play
in shaping individuals cannot be overemphasized in the case of the new
math. It was widely assumed at midcentury that learning math counted
as learning certain mental habits, which might in turn structure a student's later actions (even far beyond mathematical fields). The period in
which the arguments of Panofsky, Foucault, and Bourdieu initially flourished, from the late 1940s to the 1970s, encompassed the development of
the new math, and was indeed a period in which the relationships among
mental habits, intellectual training, and social order were explicitly and
intensely debated. The new math was not like a technology of the self.
The new math *was* such a technology.

Midcentury Schools

The new math drew on this long legacy of math as intellectual training. It nevertheless bore the imprint of specific midcentury American
forces. Mathematics in the twentieth century proceeded against a particular set of political, social, and cultural assumptions. There are many
strands to be unraveled in subsequent chapters, but two are important
at the outset: the role of math as part of a liberal education and the sig-

nificance of twentieth-century developments in mathematics and the sciences generally.

First, one key distinction between the midcentury reforms and almost all earlier attempts to teach mathematics as mental discipline was that by the 1950s the subject had been folded into the idea of universal education within the United States. For much of recorded history, formal education or "schooling" was reserved for the few. Over the course of the twentieth century, however, nation-states increasingly required ten or more years of education as a foundation of democratic social order. Schools were configured to provide students the basic skills needed to exercise the rights and responsibilities of citizenship. The intellectual training of citizens did not primarily mean memorizing the process of lawmaking or the names of state capitals but learning how to reasonably and responsibly exercise one's civic duties.

Even within this broader expansion, the situation in American education was exceptional. By midcentury, for example, only a third of British sixteen- to seventeen-year-olds remained in school; in the United States only a third of this age group was *not* in school. Few other countries took on the burden of educating such a high percentage of citizens. The American case was primarily the result of the decision to include high school education as part of general education rather than as specific training for the minority who would go on to college.[20] (Many factors were involved in this expansion, of course, from education geared to "Americanize" recent immigrants to the replacement of child labor practices with mandatory school attendance.) By the twentieth century, studying mathematics served a purpose other than cultivating gentlemen or training surveyors, accountants, astronomers, and those in similar technical fields. Teaching math was a part of the political and intellectual machinery intended to construct a modern democratic society.

That is not to say the new math created a new conception of the "citizen" or of the "rational self" during this period. The mathematicians and teachers involved certainly did not formally theorize the relationship between their work and overarching conceptions of citizenship or rationality in any extensive way. By and large, they presumed that there was a mostly uniform set of skills and habits that American citizens ought to share, not recognizing ongoing struggles to define and limit just who counted as a "citizen."[21] Any transformation in the new math era was rather in the understanding of the role elite academic disciplines might play in shaping the habits of citizens, particularly under the aegis of the

Cold War. A midcentury influx of money and urgency led to increased investment in the sciences, which subsequently produced new tools for understanding and shaping society. In many disciplines, a notion of "Cold War rationality" emerged as a semicoherent way of thought. At the same time, it became increasingly credible that different sociopolitical orders mapped onto (if not resulted from) different personality types, and social scientific research on the "American self" or the "Soviet personality" flourished among these efforts to identify consequential distinctions. In the context of this new role for academic disciplines in defining and shaping national characters, SMSG's mathematicians posited elite mathematical practices as a desirable component of American intellectual training.[22]

Midcentury math curriculum reforms were not limited to the United States. Mathematicians and teachers around the world worked to write new textbooks and pedagogical materials in the 1950s and 1960s, with particular success in Europe. The American textbooks themselves had international influence, after SMSG translated its books into Spanish and then marketed them in Central and South America, in conjunction with teacher-training institutes sponsored by the United States.[23] The American experience of reform was unique, however. The country had an expansive system of public education with neither a central educational authority nor national standards for textbooks, curricula, teacher training, or school administration. Central bureaucratic committees or expert panels could not simply mandate curriculum changes in American schools as they could in France, for example. Neither mathematicians nor politicians could decree the content to be learned in math classrooms. Any large-scale curricular changes in the United States required action on the part of a very large number of parents, teachers, textbook writers, principals, and school boards, often with conflicting perspectives and interests. As a result the reforms had to be repeatedly sold to the public, introducing a level of political wrangling and opposition uncharacteristic of many other national settings.

The experience of midcentury curriculum reform in the United States is, consequently, a good place to look for the many layers of negotiation and contingency surrounding the meaning, purpose, and nature of mathematics education. Of course, local communities have long used schools and the curriculum as venues for debates about social and political order—as historian Carl Kaestle wrote over thirty years ago about nineteenth-century "common schools": "Moral education thus overlapped citizenship."[24] In the 1950s, however, schools were incorporated for the first time into *national* curriculum-based reform efforts. Cold Warriors, integrationists, and antipoverty advocates all included educational reform as part of their larger platforms. American public schools were both sites of and mechanisms for reform. Those wishing to change society sought to change the schools.

American educational institutions are fit topics for political and cultural history because schools have long facilitated political engagement. Whether lamenting the inability of the schools to effect social change, contesting school board elections, or complaining about the curriculum and teachers, American often have their first taste of political activism once their children enter school. During the political ferment of the 1960s and 1970s, Americans saw schools as a place for registering and responding to larger national initiatives. Parents joined together in opposition to segregated schools and then later against forced busing. Advocates for racial minorities and women denounced traditional barriers to educational institutions and promoted more diverse textbook representations. Christians pushed for prayers to be introduced into public schools and sex education to be kept out. Urban community leaders promised adequate schools for all, while parents of disabled children fought for equal access to public education. Americans contextualized the new math within this matrix of intellectual and moral commitments.

The New Math = Novel Mathematics?

The second distinguishing feature of midcentury reform was a renewed emphasis on the role of mathematics itself. SMSG was motivated by a belief that the nature of mathematical knowledge had fundamentally changed in the twentieth century. Despite many differences between the new math groups, one generally accepted axiom was that math textbooks' and teachers' traditional reliance on memorization and regurgitation gave students a misleading sense of what mathematicians *do* and what mathematics was *about*. Many proponents of the new math wanted to emphasize conceptual understanding of the practice of mathematics rather than rote learning of particular facts and skills. This was not usually envisioned as a zero-sum game: reformers believed that computational ability would ultimately improve if students understood the systematic nature of the algorithms and symbols they were employing. The

new math would be recognizable by a novel classroom vocabulary—
especially that of sets—programs encouraged in order to help students
understand the underlying structure of arithmetic algorithms and algebraic techniques. SMSG's writers and defenders pointed to the history of
mathematics over the previous century to claim that only *now* did they
know the subject's true nature.

A school mathematics program, SMSG claimed, should not neglect
ancient or outmoded mathematics but should fit traditional ideas into
the "total picture" of mathematics that practitioners currently held.[25]
That picture was one of the "basic unity" of mathematics' methods and
concepts, with each discrete topic exhibiting the structure characteristic
of the discipline as a whole. "Structure" was meant both in the sense that
mathematics was the study of well-defined objects and properties and in
the sense that mathematics was a highly systematic discipline, more concerned with methods of reasoning than with any particular set of objects
or facts. These structures and systems distinguished SMSG's math for
"intelligent citizenship" from math as an exercise in rote memorization.

Like curriculum reformers across the sciences, new math's proponents
emphasized that in a technological age, an age when wars were won and
domestic tranquility secured through the fruits of scientific achievement,
students needed to know mathematics and the sciences as much more
than a set of facts.[26] The history of the new math intersects many aspects
of this new Cold War understanding of the sciences—the geopolitical
realignments brought about by nuclear weaponry, claimed manpower
"shortages" and armament "gaps," increased mathematization of many
fields, the military-industrial-academic complex, among others.

The claim that the new math entailed a new conception of mathematical knowledge, however, does not imply the introduction of a wholly different set of topics into the schools. Just a cursory glance at an SMSG
textbook table of contents might lead to the conclusion that SMSG's version of new math was not new at all. The same topics were there, largely
in the same order and configuration. Arithmetic was still the primary
subject for the first seven or eight grades, followed by yearlong courses
in algebra and geometry, and then a year with functions, or some other
advanced topic for the few students who took additional math classes.
Even the "new" concepts introduced were typically products of late
nineteenth- and early twentieth-century mathematics research, hardly
cutting-edge. The key novelty, Begle insisted, was taking "a new point of
view toward old topics."[27] As Begle later explained, "the essential char-

acteristic of a modern mathematics program is that it endeavors to make sure the student understands mathematical ideas and the way they fit together."[28]

Rather than simply noting the distinction between "understanding" and "rote learning," however, this book explores the fundamental instability and ambiguity of what it means to "understand" mathematics. No one claimed that students should not be competent calculators, and, in any case, one can perfectly well understand mathematics as a set of algorithms and facts. What counts as understanding is flexible: mathematics can be used to improve memory, to develop creative problem-solving heuristics, or to hone the skills of logical deduction. SMSG's emphasis on understanding was based on a particular conception of the connection between the nature of mathematical knowledge and the intellectual needs of midcentury Americans.

SMSG and the Context of Reform

Although its scope alone might have given SMSG credibility and prominence among midcentury reform efforts, the group also enjoyed "official" status. The organization was entirely supported by federal funds and was directed by professional mathematicians working alongside teachers and professors of education. SMSG was widely understood as the premier organization for mathematics reform, with its products endorsed by professional mathematicians.

Nevertheless, SMSG was just one organization among many. Over a dozen significant reform groups operated during SMSG's existence and some, particularly the College Entrance Examination Board's Commission on Mathematics, had previously publicized plans for an entirely new high school curriculum.[29] One of the earliest mathematics reform efforts was started at the University of Illinois in 1952 when a professor of education, Max Beberman, founded the University of Illinois Committee on School Mathematics. The Illinois Committee began as a collaborative effort between the school of engineering, the school of education, and the university laboratory high school to devise a course of secondary school mathematics. The goal was for engineering students to be able to take calculus far earlier in their college careers.[30] Eventually, the group expanded into elementary initiatives as well. Other colleges and institutions had their own projects, with very different approaches from either

Illinois's efforts or what SMSG would become.[31] These predecessors provided SMSG with material, models, and experienced writers. Begle assured SMSG staff that they would be free to design SMSG's texts as they saw fit, but he used the ongoing curricular work of other groups as a starting point and looked to hire participants from previous efforts.

These other groups are indeed important for understanding the entire scope of midcentury reform but, partly for reasons of clarity, are treated peripherally in this account. Beberman himself, to take only one example, had a sometimes-contentious relationship with SMSG—but SMSG also assimilated and built upon his work. Furthermore, unlike some contemporary groups, SMSG considered its primary emphasis to be portraying the nature of mathematics correctly. From the start, SMSG was an organization of mathematicians, its products understood to be mathematically rigorous, and its interventions primarily based on the importance of getting the math right, not the pedagogy right.

SMSG consequently produced one of the more conservative reform efforts. Its leadership rejected recommendations from groups like the Cambridge Conference on School Mathematics to include far more sophisticated mathematical ideas in the curriculum.[32] Nor did SMSG challenge the idea that math class would have a teacher in front of mostly silent students, using a textbook for examples and reinforcement. Examinations and other traditional forms of evaluation were maintained; classroom practices continued to emphasize lectures and worked examples. SMSG pushed for few institutional changes, focusing instead on the role of new textbooks. SMSG's most original pedagogical innovation might have been to allow students to write in their textbooks and keep them, a recommendation generally ignored given the financial constraints of most school districts. SMSG largely downplayed "discovery methods" prevalent in other groups, which emphasized the need for students to figure out relationships, algorithms, and rules themselves. One of the motivating factors of starting SMSG was to ensure that more radical reforms did not become the model texts by default, and instead to propose something that most schools would be likely and able to use. Looking back on SMSG's involvement, Begle later explained that while the organization was successful in changing the content of textbooks, "SMSG was an observer rather than a participant in pedagogical innovations."[33]

The true novelty, and significance, of SMSG's textbooks ultimately lay not in the introduction of new concepts or pedagogical techniques but in their intended effect on students' minds. SMSG's textbooks were innovative—and controversial—in the assumptions authors made about the nature of mathematics and the virtues studying mathematics instilled in students. SMSG's authors were motivated by the desire to teach "modern" mathematics, but the point of the reforms was to cultivate specific mental habits and thereby shape the way students thought.

Operating in the context of many curricular reforms, SMSG took the disparate efforts and made them into a credible, expansive, and consequential set of model textbooks. SMSG was not simply synonymous with the new math, however. In particular, the experience of new math in elementary schools was only partially influenced by SMSG itself. SMSG therefore played a central role, but ultimately the contours of its organizational history did not exactly follow the same lines as the overall story of the new math. The slippery and complicated idea of *the new math* was a concept made manifest in SMSG's textbooks but also was an entity all its own.

Organization of the Book

The New Math follows a generally chronological order, but each chapter focuses on a different aspect of the curricular reforms. This allows the rise and fall of the new math to be understood through the voices of the diverse groups who spoke out on the reforms at various moments. It also intertwines educational history with American intellectual and cultural history, and the history of science with precollegiate curriculum history. Rather than dwelling on the relevance or meaningfulness of these traditional distinctions, the book simply treats the school curriculum as a strategic site for examining the nature of mathematical knowledge and the role of mathematical knowledge in the intellectual training of citizens. There is no reason to separate the "mathematical" pieces from the "curricular" or "historical" pieces. Decisions about how to teach mathematics counted as decisions about the nature of the discipline and the disciplining value of mathematics.

The next chapter analyzes the bureaucratic and financial origins of federal support for curriculum development in general and SMSG's curricular projects in particular. The origins of the new math are primarily located in the push to fight the "Cold War of the classrooms." Long before the 1957 launch of Sputnik, critics had called for an overhaul of the American education system, and the Soviet satellite provided an opportunity ripe for exploitation. The use of classrooms as a site of Cold War politics was neither inevitable nor self-evident. While some reformers pointed to the deficiencies of "progressive education" or the need for investment in infrastructure, the most compelling push for educational reform involved the role mathematics could play in fostering mental habits. Politicians supporting curricular reform claimed that the crises facing the nation were ultimately rooted in problems of intellectual discipline. Congress's shift to support broad federal education legislation grew out of the conclusion that the NSF's curricular initiatives would provide students the scientific background required for an increasingly complex world and reintroduce rigorous intellectual training into schools.

The third chapter introduces SMSG's textbooks, first focusing on the perspective of the mathematicians who designed them. These mathematicians promoted a controversial conception of the nature of the subject in junior and senior high school textbooks, a version of mathematics as abstract, structural knowledge. This vision was not universally shared, and the chapter examines divisions within the community of professional mathematicians. These debates centered on diverging views of the Bourbaki program for mathematics, and especially on disagreements over the relationship between the nature of mathematical knowledge and the field's many applications and uses. Mathematicians cared deeply about how textbooks portrayed their subject largely because these debates were never only about school mathematics. Behind claims about the role of mathematics in shaping intellectual discipline was a specific defense of the importance of learning "modern" mathematics. Descriptions of what mathematics *was* and *why* it should be learned involved claims about the broader relationship of mathematics to intellectual discipline.

The fourth chapter analyzes the elementary textbooks and claims made about how arithmetic should be taught. During an era when the characteristics of "free" minds were opposed to those of "Soviet" and "closed" minds, the goal of learning mathematics was not simply the acquisition of arithmetical facts. The ability to compute implied the ability of students to acquire certain and reliable knowledge. SMSG's elementary textbooks show that the philosophy behind the reform of elementary mathematics was substantively different from that of the secondary textbooks: no one claimed that elementary schools were somehow failing in their primary goal of cultivating arithmetical ability in students. Rather, SMSG's writers believed that traditional approaches were training students to calculate in the wrong way. The now-infamous introduction of set theoretic notation into the schools emerged from an attempt to refashion the ideology of elementary arithmetic. Students who learned math in this way would learn how to acquire reliable knowledge.

While chapters 3 and 4 primarily track one specific curriculum and the claims made about it and in its textbooks, the next two chapters focus on how the broader ideas of the new math were deployed and received. In chapter 5, the book shifts focus from the imagined role of textbooks in intellectual training to the very real movement of actual textbooks from the drawing boards of university mathematicians to the desks of American schoolrooms in the mid-1960s. SMSG used legions of testing sites, rapid redrafting of preliminary editions, and close connections to commercial publishers to overcome the traditional barriers to large-scale reform. Shifting away from the ideas embedded in specific textbooks, this chapter uses educational journals, textbook publishing records, and school board archives to examine how the new math actually revolutionized the content of American mathematics education.

The process in high schools was substantively different from that in elementary schools, a fact of great consequence for teachers' and parents' understanding and acceptance of the curriculum. SMSG's close collaboration with secondary school teachers proved to be an especially successful method of effecting educational reform. The organization played a much-diminished role in the more lucrative elementary textbook market, however, and was far less successful in incorporating the input of elementary teachers. Such distinctions were not widely noted at the time because in both settings the *justification* for reform was uniformly based on the supposed importance of "modern" math for the shaping of "modern" students. The new textbooks—with their claims of mathematical authority—received broad (and politically bipartisan) support for their mission to train citizens prepared to meet the challenges of a rapidly changing world.

The final chapter shows how detractors of the new math followed the curriculum's promoters in eliding distinctions between politics, mathematics, and pedagogy. The new math was not rejected primarily because of falling test scores or because students were struggling to learn to add, but because students were failing to be disciplined in the right way. Parents, teachers, and other concerned citizens began to oppose the new math on the basis of an alternative view of how students should be trained, one that emphasized functional goals and was rooted in rote drills. This chapter ties the rise of the "back to basics" movement to the resurgence of conservatism more broadly. Schools and the curriculum were central sites of political mobilization, particularly because they were understood to be places where cherished "values" were under siege. Inextricable from the desire for this transformation in moral authority was the claim that math class should emphasize "traditional" methods of reason—facts and basic skills—rather than the systematic methods of reasoning valued by SMSG's mathematicians.

The New Math ends with a brief account of the trajectory of math education after the demise of the new math. Many issues remained the same—should math teachers emphasize facts or processes, memorization or creativity?—and although no longer labeled as such, many textbooks retained innovations introduced in the new math era. Rather than lamenting the cyclical and seemingly intractable problems of American mathematics education, these problems might be better understood by critically contextualizing them as products of a particular political history.

The epilogue also argues that the school curriculum, when properly historicized, grounds and interrogates otherwise abstract or vague discussions about political change. In the postwar era, the curriculum served as a place through which, on which, and about which questions of social order, moral authority, and political power were negotiated and debated. Parents cared deeply about what their children learned, and how their children were taught to think.

The math curriculum, and the new math in particular, is certainly not the only midcentury subject where conceptions of reasoning well were engaged and contested. Controversies about how to teach reading, for example, entailed both psychological models of knowledge acquisition and commitments to promote mental discipline. Heated quarrels about "phonics," "look say," and "whole language" approaches also raged in this period—with similar camps of "progressives," "disciplinary experts," and "back to basics" proponents.[34] It was no coincidence that the most famous of the anti-new-math polemics, *Why Johnny Can't Add* (1973), played on the title of Rudolph Flesch's *Why Johnny Can't Read* (1955).[35] In both cases, pedagogical choices entailed ideological commitments.

Even so, literacy debates never quite implicated the vocabulary of epistemology or morality that mathematics did. Learning to glean meaning from words on a page was not the same as learning to move from postulates to reliable conclusions; debates about the most efficacious road to literacy became largely irrelevant once students could read adequately. It was not so in mathematics: at every level of instruction, teaching mathematics has had a much stronger connection with intellectual rigor and discipline. The only other traditional fields for the inculcation of intellectual discipline, grammar and rhetoric, largely ceased to be used in this way by the twentieth century, while mathematics has retained the association. Perhaps more than in any other modern subject, the design and evaluation of the math curriculum involved decisions about the desirability of particular mental habits.

The "American Subject" of this chapter's title is therefore intentionally ambiguous—referring both to mathematics and to the students whose minds were meant to be trained by the math curriculum. Descriptions of the nature of mathematical practices were also descriptions of forms of reasoning; debates about pedagogical choices within textbooks were also debates about the sort of reasoning applicable most widely and virtuously outside the classroom. Evaluation of the math curriculum turned on the evaluations of intellectual, social, and political order.

CHAPTER TWO

The Subject and the State

The Origins of the New Math

The new math was an unlikely project. Few mathematicians spend much time thinking about how to portray algebra to ninth graders, and most Americans have something other than textbooks in mind when they pay their taxes. The pedestrian nature of current education legislation would have been mystifying to midcentury Americans; broad federal funding of any sort of educational initiative was then extraordinarily rare. Writing a set of textbooks hardly guarantees they will be noticed or used. At the time, the new math seemed more likely to end up an unfulfilled fantasy.

Yet, SMSG received millions of federal dollars to change the way children learned mathematics. This was made possible by the success of claims that the United States risked losing the "Cold War of the classrooms" to the Soviet Union.[1] Throughout the 1950s, many scientists and politicians worried that the technological knowledge required for military success would never be developed if schools failed to train enough scientifically literate students. Both supporters and critics of American education found such causes compelling. Longtime educators hailed the emphasis placed on improving schools. Critics of the status quo praised the replacement of educators by scientists as textbook authors. The new math was a project of academic and political elites, intended to effect broad, top-down, educational reform. The curriculum was born out of the confluence of scientists, educational critics, politicians, and professional educators opportunistically looking to the schools as weapons for waging the Cold War.

The new math was not, however, solely a response to Cold War crises, and in particular was not caused by the Soviet launch of Sputnik into orbit in October 1957. For one thing, small reform efforts had long been under way, and the National Science Foundation had indicated an interest in a large-scale mathematics curriculum project prior to the launch. Moreover, there were no immediate educational consequences of Sputnik. It was, after all, just an 184-pound orbiting radio. As historian Lawrence Cremin remarked, after Sputnik "the public blamed the schools, not realizing that the only thing that had been proved, as the quip went at the time, was that their German scientists had gotten ahead of our German scientists."[2] The launch was hardly secret, and both the Soviet and U.S. governments had planned satellite launches for the International Geophysical Year in 1957–58. Although Sputnik's beeps had evident military overtones—the Russians had successfully launched intercontinental ballistic missiles just months earlier, and Sputnik seemed to confirm their mastery of rockets powerful enough to reach American soil—the main effects of the launch might have been a reshuffling of military priorities. President Dwight Eisenhower had specifically separated the military and scientific aspects of U.S. missile work, hoping to claim the skies for peaceful purposes, a goal Soviets themselves might have fostered with the launching of an innocuous device.[3] Cold Warriors might have simply advocated increased military budgets or intensified diplomatic efforts after the launch; scientists might have directed money to research institutes or professional training programs rather than to the development of school textbooks.

Scientists and politicians had to make the case that geopolitical exigencies like Sputnik demanded action at the level of the math curriculum. A more telling explanation for the push to develop new math textbooks can be found in the persistent belief that schools had failed to provide appropriate intellectual training. The Cold War of the classrooms was not primarily fought over teacher preparation, classroom size, laboratory provisions, or even graduation requirements. It was fought over the relationship between the subjects taught in the classroom and the intellectual discipline of the nation's students. While there was a specific role for mathematics, the impetus for reform also affected other fields, in particular physics and biology.[4] This chapter outlines the apparent intellectual deficiencies that were used to justify the new math's development; subsequent chapters explicate precisely how SMSG positioned its mathematics textbooks to cultivate particular intellectual virtues. The Cold War of the classrooms—and the new math it begot—was ultimately a product of faith in the ability of a well-designed curriculum to discipline American minds.

Skills

Many considerations made a federal effort to reform American schools unlikely. There was first the issue of whether schools *should* be reformed. While data are hard to come by, the evidence that exists suggests midcentury parents were happy with the quality of education provided. Early postwar attempts to improve schools emphasized cooperation with educational authorities and gradual rather than radical change. In 1955, the National Opinion Research Center, for example, reported that 82 percent of Americans were satisfied with the education their children were receiving. Arthur Zilversmit's close study of Illinois schools in the 1950s showed that few citizens had any complaints about the schools. Such formal and informal surveys can be misleading, as approval ratings often track the percentage of citizens with children: parents usually rate schools more highly than nonparents. Nevertheless, there is little evidence of widespread dissatisfaction. [5]

Moreover, American schools were decentralized. States or individual districts specified the content, hired the employees, and funded the schools through tax receipts. Textbooks were rarely updated—few districts had the money required to purchase new materials regularly. Schools remained local and often small: at midcentury nearly 60,000 one-teacher schools were still scattered across the country, and even years later many students attended high schools with populations of less than 300, and therefore with little variety or choice in courses.[6] The federal government could not mandate that students take mathematics, teachers learn more mathematics, or districts adopt more sophisticated mathematics texts.

Furthermore, there was no clear or necessary role for the federal government in education policy. Given the U.S. Constitution's relative silence on matters of education, the federal government had been only episodically involved in school legislation, including the 1787 North West Ordinance, nineteenth-century land grant universities and agriculture experiment stations, the 1917 Smith-Hughes Act for vocational education, Great Depression educational initiatives, and the Servicemen's Readjustment Act of 1944. In the decades prior to the late 1950s, however, legislation providing *general* aid to education (rather than, say, aid for veterans, agriculture research, or other special groups) had almost no chance of success in Congress.[7]

By the 1950s, fundamental disagreements among supporters of parochial schools, pro-integrationist legislators, and Cold Warriors ensured that those pushing a more active involvement of the government in education faced an uncertain future. Political scientist Stephen K. Bailey described the opposition as based on the three Rs of "race, religion, and reds." Southern segregationists would join with northern supporters of Catholic schools to block aid to public schools generally. Eisenhower only reluctantly—and then tepidly—supported any sort of broad federal education legislation during his two terms.[8] Given such entrenched opposition, it was by no means taken for granted that an appropriate response to Sputnik would be federal involvement in the schools.

Despite these obstacles, schools were an increasingly common site of national reform efforts. Schools had always been mechanisms through which localities aimed to inculcate particular values and virtues into students, and occasionally local disputes like the Scopes trial became nationally resonant. After midcentury, Americans increasingly looked to the schools. As historian William Reese has argued, "between the end of World War II and the ascendancy of Ronald Reagan as president, no institution bore a heavier burden for improving society than America's public schools."[9] Schools became both sites of and mechanisms for reform.

Battles about the nature and direction of American civic life at midcentury often centered on control of and access to schools. Famously, the lawyers for the civil rights movement chose the schools as the site for their legal claims. If public trains had been the institution that established the doctrine of separate but equal, the public school system would be the site of its demise.[10] From school prayer to sex education, parents fought over the appropriate extent of extracurricular education. As colleges and universities ended (or eased) admissions quotas and opened their doors to women and minorities, schools offered the ticket for upward social mobility. Organizations like the College Board, originally a college placement service for boarding school men, began to help universities locate and recruit talented students from diverse backgrounds. In the effort to promote "excellence" in American schools, access to elite educational, cultural, and political institutions increasingly—although never, of course, completely—depended on test scores and academic performance rather than family connections. John Gardner's influential 1961 book *Excellence: Can We Be Equal and Excellent Too?* examined these tensions. Gardner, president of the Carnegie Corporation and later secretary of the Department of Health, Education and Welfare, attempted to specify how the nation could continue to emphasize broad measures of equality while also promoting individual merit. He concluded that the successful maintenance of American democracy required schools to develop intellectual abilities in students. In postwar America, schools were widely conceived as a crucible for the reformation of society.[11]

Scientists echoed reformers like Gardner in describing schools as weapons in the Cold War. The theme was repeated ad nauseam in legislative hearings when the then senator Lyndon Johnson convened his Preparedness Subcommittee of the Committee on Armed Services weeks after Sputnik—a group more familiar with the technical details of missile design than with school curricula. Engineer Vannevar Bush, the former director of the office of Scientific Research and Development and an influential architect of postwar arrangements between scientists and the government, complained that the American educational system had not adequately promoted science. Educators had become complacent, Bush alleged, allowing students to take easier and less-demanding subjects. He warned that the country needed to "wake up to the fact that we are in a tough, competitive race where we have got to do a lot of good tough work, and that that begins just as soon as the youngster goes to school." Only when we got students to "struggle with tough subjects" might the nation be able to find and train scientists and engineers for the future. Air Force general James H. Doolittle, chairman of the Scientific Advisory Board, connected the dots explicitly by explaining that the "foundation upon which their [the Soviet Union's] new technology is built is their excellent educational system." Soviet students, he claimed, work harder, take more advanced classes, and study for longer periods each day. Doolittle predicted that if American educators continued to give students the freedom to avoid difficult courses, the nation's technological supremacy would surely falter.[12]

Even before the launch of Sputnik, military brass had asserted that educational deficiencies would have grave implications for the Cold War. Admiral H. G. Rickover repeatedly warned that the lack of intellectual discipline in schools entailed not just a moral or social problem but also a problem of international security. Spurred on by the threat of Soviet scientific prowess, Rickover suggested that nothing less than the future of the free world was at stake, and that he was impelled to speak out because of the possibility that the USSR would soon overtake the United States in the Cold War: "The future belongs to the best-educated nation. Let it be ours."[13] Both Rickover and Doolittle used the idea of rigor to emphasize that military and national strength emerged from quality education. Tough courses made tough students, able to win the Cold War.

After Sputnik, Rickover's warnings seemed prescient. Arkansas senator William Fulbright argued that "the heart of the contest with the Soviet Union is education," while New York representative Herbert Zelenko predicted that in the future, "defense is no longer a matter of muscles and masses. . . . Formulas and equations have taken the place of spears and guns." He concluded that "education is the true defense." Alabama senator Lister Hill repeated Rickover's urgency in pushing federal involvement with the schools: "intellectual discipline is essential to our national purpose." As Hill would exhort his fellow congressmen, "We Americans know we must mobilize our Nation's brainpower in the struggle for survival."[14] The message had become a mantra: schools were critical to winning the Cold War.

The politicians pushing federal investment in education did not point to crumbling buildings, or out-of-date textbooks, or even teacher certification requirements. The Cold War of the classrooms involved proper intellectual foundations. Former president Herbert Hoover lamented that contemporary students did not work hard enough, and Senator Ralph Flanders of Vermont regretted in hearings on education that courses were being offered in dance and skiing instead of basic subjects. As California representative John Phillips remarked in congressional hearings, funding scientific research without improving the schools was like putting "delightful icing on a cake that is made with very poor flour." The issue of scientific manpower was at heart a concern of the schools, because in them "there is no training in mental discipline. Well, there is no training in any kind of discipline that I have observed." Montana senator Mike Mansfield similarly complained that "there is inadequate training and discipline in high schools." New York's Ralph Waldo Gwinn lamented the decline in fundamental courses, "in favor of progressive educational material, such as social adjustment, sociology, home economics, bird watching, nature study, field trips, and so forth."[15] It was not just a matter of increased funding or better teachers. Students' minds were simply not being trained in the right way.

Progressive Education and Its Critics

When Gwinn used the language of "progressive educational material" he was pointedly referring to the influence of progressive educators. Critics in Congress and university professors overwhelmingly blamed the influence of progressive educators for the "decay of our educational standards," in the words of Senator Flanders.[16] Progressive education was a diffuse movement linked philosophically to the early twentieth-century writings of John Dewey, who attempted to distinguish education from other fields and establish it as a scientific discipline in its own right. In addition to incorporating new philosophical and psychological research into pedagogical and administrative practices, progressives envisioned an expanded role for schools in the formation of democratic attitudes, following Dewey's emphasis on the potentially transformative power of schools in the public sphere.[17]

Progressive education was driven by transformations in twentieth-century education generally. Increasingly, teachers were identified as female, unionized professionals, trained in official teacher-preparation programs. Progressives were closely associated with schools of education, especially Teachers College in New York, and with the growth of professional education research during this period. The first decades of the twentieth century saw the emergence of powerful central educational authorities, connected to the increasing use of high schools, and to rising attendance generally. These authorities were themselves often products of schools of education and filled open positions with like-minded peers in the 1930s and 1940s.[18] Thus, when Flanders complained about progressives, he made certain to blame not only Dewey's "morally weak" philosophy but also the influence of Teachers College.[19] The problem was that education had been hijacked from disciplinary experts by "educationists" who claimed to be knowledgeable about pedagogy, not mathematics, literature, or the content of any other particular field of knowledge.

Those who defended progressives publicly in this period often did so by pointing to the changing role of the schools in the first half of the twentieth century. Progressive education emerged as schools increasingly bore the burden of training all students, the majority of whom did not plan to attend college. Progressives, particularly followers of Dewey's work at the University of Chicago, were focused on addressing

the concerns of urbanization and immigration through the incorporation of scientifically trained experts and the expansion of the psychological sciences. Early twentieth-century schools became part of a broader
economic transition, with college seen as just one of many choices after
graduation. Harvard's president James Bryant Conant concluded in the
late 1950s that the modernization of education was as much a product of
child labor laws as of the writings of Dewey.[20] By midcentury, American
schools and teachers could no longer focus on the capable few. Latin was
taught next door to shop class, as educators had to account for students
of varying interests and abilities.

Progressives promoted a diverse range of initiatives and innovations
in response to the changing school population, including mental hygiene,
home economics, project-based learning, child-centric classrooms, and
early guidance counseling and testing. While some progressives argued
for a shift away from traditional subjects to more "practical" ones, others simply promoted new pedagogical techniques. At the end of World
War II, despite little documentation on how far progressive practices
had actually influenced schools or how coherent progressive tenants
were, enemies of progressive education could describe it as the "conventional wisdom." If so, it was a short-lived triumph, and by the launch of
Sputnik the Progressive Education Association had quietly disbanded
from lack of support.[21]

Despite the Progressive Education Association's rapid decline in the
1950s, critics continued to blame its philosophy and practitioners for educational problems. Teachers, one common argument suggested, were
more focused on "life-adjustment" education—preparing children to fit
in—than on teaching students the intellectual skills to understand and
change the world. Robert Hutchins, former president of the University
of Chicago and frequent critic of midcentury schools, complained that
progressives' ideas of life adjustment risked grave consequences: "At
least during a cold war, the doctrine of adaptation leads remorselessly
to indoctrination." Hutchins maintained that schools needed to cultivate
students' "intellectual powers" in order to ensure the "fullest development of the nature of man." Intellectual powers emerged in a liberal education, Hutchins suggested, only from the study of "ideas that have ani-
mated mankind." Schools should promote careful study of the academic
disciplines—"familiarity with the best models"—not the watered-down
formulations that all too often made their way into textbooks.[22]

One of the most influential criticisms came from someone who was

himself a product of progressive schools, Arthur Bestor, whose most influential work was his 1953 *Educational Wastelands*.[23] Often cited as a
critic of progressive education and professional educators, Bestor located the problems of education not in the basic philosophy behind progressive ideals, but rather in the diminished emphasis on intellectual
rigor. The problems with the schools, he claimed, were problems with
the way in which educators trained the minds of students. Teachers were
overly concerned with the immediate needs of students and had moved
too far from the disciplines of intellectual inquiry. A professor at the
University of Illinois, Bestor believed the proper aim of schools was the
inculcation of "disciplined intelligence," and he meant both discipline
as regulation of body and mind, and discipline as a formal mode of inquiry. Educators, especially those who claimed the mantle of progressivism, had rechristened the disciplines as "subject-matter fields." For
Bestor, such a distinction indeed made a difference: mathematics, biology, and history were known as disciplines precisely because they were
meant as ordered ways of thinking. The practices and the content of the
disciplines provided each student "intellectual mastery and hence practical power over the various problems that confront him." Schools had
failed most students by providing this sort of rigorous intellectual training only to the few, and Bestor helped create the Council for Basic Education to promote the teaching of the traditional disciplines to all students.[24] Schools should equip students for life's problems not by focusing
on practical matters but rather by preparing students intellectually for
the challenges of contemporary life.

Both Bestor and Hutchins cited the dangers of McCarthyism and
of mindless "Red Scares" as evidence for the importance of disciplined, liberal education. Hutchins denounced loyalty oaths for teachers as a product of Senator Joseph McCarthy's "irresponsible fulmina-
tions" and warned against the tendency toward a "flat conformity of
life and thought." A liberal education was critical to a free society since
the "only serious doubt" that existed about democracy was precisely
"whether it is possible to combine the rule of the majority with that independence of character, thought, and conduct which the progress of any
society requires."[25] Bestor likewise defended the freedom of inquiry and
teaching. He believed a liberal education would model and instill support for a free system of government.[26] The most important target of the
McCarthyite crusade—outside of Hollywood—was the public school system. Teachers came under siege as potential subversives, and towns like

Pasadena, California, erupted over the hiring of progressives and the introduction of "subversive" textbooks. Intellectuals debated whether communist sympathies immediately disqualified a person from disinterested academic inquiry. Anticommunists had conflicting ideas about how to counteract Soviet influence but were in general agreement that a liberal education was critical not only for checking the demagoguery of men such as McCarthy but also for guarding against any actual communist threat.[27]

Academics commonly proposed their own methods as a model for responding to these midcentury challenges. Professors' reasoning skills were held up as an antidote to the authoritarian impulses of the masses. As historian David Hollinger has noted, academic science was a "magnificent ideological resource" for intellectuals seeking to define the nature—and relative value—of public or private knowledge. Psychology and sociology departments attempted to measure and understand the risks of "mass society" and the concomitant threats to individual autonomy.[28] Richard Hofstadter's 1963 survey of "anti-intellectualism" warned that American educators had gradually become estranged from academia, with low-status teachers and their professional organizations too focused on promoting progressive pedagogy and preparing students for participation in mass society. "Anti-intellectualism," Hofstadter warned, naturally followed. In the midst of a general survey of curriculum reform, Sterling McMurrin, a professor at the University of Utah, explained anti-intellectualism as a product of "our failure to value knowledge for its own sake as well as for its uses, to fully respect reason and evidence; it is our willingness to be controlled by passion and emotion and by hearsay and propaganda rather than to make our decisions and determine our actions by the more sure devices of knowledge and disciplined intelligence."[29] "Disciplined intelligence" was often little more than a code word for the sort of intelligence professors possessed.

Criticism of education came in many flavors at midcentury, from those decrying the influence of progressives to others lamenting the decline in "tough" courses. Although neither Bestor nor Hutchins was a scientist, their criticism lay the groundwork for the scientists who devised new curricula in hopes that doing so would help spread rational thought and fight anti-intellectualism. Only when mathematicians were in charge did federal money materialize, a fact not lost on teachers who had been applying fruitlessly over the years for funds to reform the math curriculum.[30] The push to have mathematicians bring more "disciplined" knowledge to the curriculum cut across political and professional lines, as reformers claimed that instilling the right intellectual discipline would solve the problems of education and help win the Cold War of the classrooms.

Civic Virtue

Those promoting discipline-specific, rigorous, intellectual training in the schools intended their work to fashion citizens better suited to the domestic and international challenges that lay ahead. One of the most cited pieces of evidence for the reformers was the 1958 report of the Rockefeller Brothers Fund, "The Pursuit of Excellence: Education and the Future of America," which warned that the educational crisis was a real one. The crisis had been made pressing by the fact that "a free society cannot commandeer talent: it must be true to its own vision of individual liberty. And yet at a time when we face problems of desperate gravity and complexity an undiscovered talent, a wasted skill, a misapplied ability is a threat to the capacity of a free people to survive." "Education," the report asserted, "is not just a mechanical process for communication to the young of certain skills and information. It springs from our most deeply rooted convictions. And if it is to have vitality both teachers and students must be infused with the values which have shaped the system."[31] Education reformers had in mind an image of a virtuous midcentury American citizen.

The close identification of virtuous citizenship and education was hardly a post-Sputnik phenomenon. Harvard's so-called Red Book of 1945 heralded the need for "General Education in a Free Society." The widely cited and influential report concluded that "the aim of education should be to prepare an individual to become an expert both in some particular vocation or art and in the general art of the free man and the citizen." If "Democracy is a *community* of free men," the report continued, then "the fruit of education is intelligence in action."[32] Similarly, a 1951 pamphlet of the National Education Association, "Education and National Security," reminded readers that schools "must educate for moral and spiritual values." "National security," the authors emphasized, was not just a matter of developing the right skills or producing enough scientists but of promoting "civic intelligence."[33] Reformers were drawing on this tradition when they described the virtues of intellectually disciplined citizens.

There was much debate about what those virtues were and ought to be. David Riesman, William Whyte, and Vance Packard all wrote about the threat of conformity at the expense of individual creativity and productivity. While corporate structures emphasized the benefits of teamwork and collaborative effort, some worried that the individual's autonomy and creativity would be lost. Much of this cultural angst centered on the risks of Soviet-style conformity—if not mentioned by name—and the need for institutions and individuals to protect against such risks. These concerns were never far from curriculum reform efforts; the head of the NSF's physics curriculum program, MIT's Jerrold Zacharias, complained that "our schools have been set up in such a way as to produce a sea of uniform mice; if you don't believe it, look at the lamps people buy, or the automobiles."[34] The right sort of intellectual training might be able to guard against the corrosive influence of cultural conformity.

Popular culture was imbued with the consequences of conformity and the subversion of virtue. *Invasion of the Body Snatchers* (1956) and *The Manchurian Candidate* (1962) warned moviegoers of the pernicious effects of seemingly undetectable traitors in the midst of the nation. Throughout the 1950s, people feared that Americans had gone soft and were unable to assert independence from authority. Sloan Wilson's *Man in a Gray Flannel Suit* (1955) gave readers a picture of the numbing and emasculating effects of corporate conformity. Ayn Rand's novel *Atlas Shrugged* (1957) warned of the creeping communitarian and socialistic tendencies which undercut the "true" capitalist, and American, spirit of autonomous creativity.[35]

Mental health experts warned of the pernicious consequences of preventing children, particularly boys, from becoming healthy, autonomous adults. Philip Wylie's *Generation of Vipers* had popularized the notion of "momism," a term that later came to stand for the variety of ways mothers coddled their children. Edward Strecker went so far as to link the smothering mother to issues of national security in his book *Their Mothers' Sons*, charging that elevated rates of psychiatric rejections and discharges in World War II were a result of bad parenting. By failing to raise independent, tough children, the nation's mothers had all but guaranteed that a generation would be unfit to live on its own.[36] Conformity and the failure to achieve autonomy were not just cultural trends but dangerous tendencies in a time of prolonged Cold War.

Schools were always at least partially at fault because they had failed to prepare students to maintain a "free" society. David Dawson of du

Pont industries explained at a national Parent-Teacher Association meeting in 1958 that "technological superiority alone will not determine the winner of the conflicts that will abound in the world of the future. They can be won only by a nation of free men—free to think for themselves, encouraged to think for themselves, and trained to think for themselves." John Gardner's book *Excellence* took much the same line. "Free societies will not survive (nor be worthy of survival) if the tradition of individual fulfillment decays from within." Gardner and Dawson claimed that by better cultivating the disciplines in schools—by training individuals to think—educational reform could fashion a free society.[37]

Discipline in the schools ensured that the freedom enjoyed by Americans would not devolve into delinquency or subversiveness. In the year of Sputnik's launch, Broadway introduced the teen gangs of *West Side Story*, including the "Officer Krupke" paean to juvenile delinquency, and Hollywood released the film *Delicate Delinquent*. At least since 1955's *Blackboard Jungle*, American popular culture had grappled with balancing virtuous individuality against the dangers of delinquency. The rise of the juvenile delinquent—and its accompanying racial and class tensions—suggested a "problem of great contemporary concern," according to one reviewer of *Blackboard Jungle*.[38]

Schools could address these concerns about delinquency by promoting mental discipline, even as they encouraged students to think freely. A widely reprinted article by University of Detroit professor Walter Kolesnik drew attention to—and lamented—the scholarly decline of the theory of mental discipline. Kolesnik argued for the continued relevancy of mental discipline as central to a school's quest to use resources efficiently in the education of students. "In short," Kolesnik wrote, "a well-trained mind would be, in its own right, the most valuable possession the school can help the student attain." Senator Strom Thurmond repeated such sentiments in Congress, noting that "no amount of money can make up for the failure to teach discipline which must be applied from without in formative years in order that it may be applied from within in mature years."[39] The right kind of discipline was imperative.

Conclusions about the interrelatedness of virtue, discipline, and the schools were not limited to academic journals, scholarly monographs, or speeches in the *Congressional Record*. *Life* magazine ran a five-part series in spring 1958 warning about the present "Crisis in Education."[40] In addition to mentioning the factors of poor teacher pay, family involvement, and the lack of programs for "gifted" children, the opening article of the series laid out the underlying problem succinctly: while the typical American student wisecracks about his ignorance and spends his weekends with friends and sports, the typical Russian student spends hours on homework and serious activities like chess. Comparing the experiences of Stephen Lapekas of Chicago and Alexei Kutzkov of Moscow led the author to conclude that the "intellectual application expected" of American students is "moderate" compared to the advanced level of the Russians. Proper discipline was both internal and external: unlike Stephen, Alexei must learn "a great deal by rote" and submit to teachers who run their classes "with a firm hand."[41] When giving examples of promising trends in American schools, *Life* pointed directly to the new curricular reforms in science and mathematics. Only with such rigorous courses could Stephen hope to catch up with Alexei.

An article in the *Life* series by Sloan Wilson—of *Gray Flannel Suit* fame—emphasized the link between a democratic society and intellectual discipline. Explaining the various factors that had led to the decline in schools in the discipline of students, Wilson concluded that the one required critical change was fostering the "virtue" of "honest respect for learning and for learned people."[42] "Democracy," he warned, "was never supposed to substitute license for discipline." In order to regain such virtues, Americans should "get tough" with schools and students, demanding changes in both behavior and the content of schoolwork. Wilson posited the schools as crucial to the ability of individuals to realize their own potential and thereby allow society to prosper. When "eager" students are labeled "queer ducks" and easy courses push out difficult ones, schools will fail to train students to think well. Wilson's article—and the entire *Life* series—highlighted the way in which the question of school reform was at heart a question about the proper means of forming citizens within a democratic society.

Scientific Manpower

The desire to reinstate rigorous mental discipline, often coupled with criticism of progressive education, formed the intellectual justification for federal support of curricular reform. The specific congressional machinations that actually produced the federal funds behind the new math, however, centered on the issue of "scientific manpower." Manpower had been a direct concern of federal science agencies since World

War II.[43] John R. Steelman, chair of President Harry Truman's Scientific Research Board, dedicated a volume of his postwar *Science and Public Policy* to the "crisis of science" as a result of the "present shortage of scientists." While pointing in part to the failure of undergraduate colleges, Steelman's committee appended a report by the Cooperative Committee on the Teaching of Science and Mathematics of the American Association for the Advancement of Science, titled "The Present Effectiveness of our Schools in the Training of Scientists." This appendix directed attention to the poor state of precollegiate education and its institutions. In the case of mathematics education, Steelman's report decried inadequate teacher training, teachers' failure to convey the meaning of arithmetical operations, students' confusion with the presentation of multiple methods, and administrators' lack of specific objectives in the elementary schools. Similarly, the report lamented that even after a 1923 study showing that the curriculum was out of date, the secondary schools had essentially failed to review or update their science courses.[44]

Despite such complaints about the role of the schools, the National Science Foundation initially focused only on graduate training in universities and research grants. After Truman signed legislation on May 10, 1950, creating the NSF, its inaugural report noted that graduate fellowships "should be the first order of business," since the training of young scientists was of "crucial importance." Crucially important, that was, for the manpower needs chronicled in the NSF's "Scientific Manpower Bulletin" and the annual summary of the nation's manpower statistics. As Henry Smyth, physicist and author of the first official accounting of the atomic bomb effort, explained, scientists were "tools of war" which should be "stockpiled" just "as we would any other essential resource." When the first appropriations were made in the next fiscal year, graduate fellowships totaled half of the NSF's entire annual budget. The government's initial answer to the manpower problem was money for graduate training.[45]

Convincing Congress yearly of the worthiness of such expenditures proved difficult. While the NSF found a sympathetic Appropriations Committee in the Senate, the House Appropriations Committee was far more skeptical about the role of the federal government in scientific research. In fact, the House voted a 98 percent reduction in the initial funding for the NSF and substantial cuts for the next few years. The typical reason was—as a House Committee member explained in 1952—that the manpower and matériel requirements of the Korean conflict made research in basic sciences "unlikely to provide assistance to the country in its immediate emergency."[46] For many politicians of the period, pure research in the sciences simply did not appear to be as compelling as direct military expenditures.

NSF's administration was forced to plead its case each year to the ranking members of the House committee, Democrat Albert Thomas of Texas and Republican John Phillips of California. Director Alan Waterman and colleagues would then head to the Senate appropriations meeting begging for rejected portions of the NSF's request to be restored. They urged senators to realize that "basic" research would eventually lead to applications and new technology, and ought at least to be supported on these grounds.[47] The NSF nevertheless had trouble securing sufficient funding in the early 1950s.

Despite these funding difficulties, the NSF tentatively experimented with improving the teaching of science in its first years. Beginning in July 1951, the NSF's board approved a proposal to sponsor "summer seminars" at selected locations to bring together practicing scientists and college teachers. NSF's administration was not initially interested in expanding this program to high school teachers, due both to the lack of funds and to the sense that having the NSF work in teacher education might "demean" the organization, while potentially conflicting with the role of schools of education. Only after additional support emerged from the Ford Foundation's Fund for the Advancement of Education did the NSF sponsor a few small-scale institutes for college teachers and one for high school teachers during the summers of 1953 and 1954. In May 1954, the consultant in charge of "Education in the Sciences" proposed expanding the NSF's purview to encompass not only teacher education but also high school and college curricula development, motivation and counseling of secondary school students, and support for talented youth.[48] Slowly, the argument that the NSF might expand into the schools gained credence. The NSF could encourage more young people to go into science, proponents promised, or at least help to develop a respect for the value of scientific work in students.

The potential value of educational work for the NSF increased substantially in the mid-1950s. On February 9, 1955, in the annual appropriations hearings, Waterman cited the preliminary findings of Nicolas DeWitt—who was conducting an extensive study comparing U.S. and Soviet scientific capabilities—to warn of the need to address training more directly. While DeWitt's research showed only modest (if any) advantages for Soviet sciences, the report caused a much bigger stir as fearmongers warned that the "scientific manpower gap" threatened the security of the United States.[49] Waterman, sensing the ability of educational initiatives to elicit more money, emphasized the potential for the NSF to expand its involvement with teacher training institutes, visiting lecturers, traveling high school libraries, and curricular reform.[50] The impetus toward pre-collegiate education only increased after Eisenhower included scientific training as a formal component of his administration's national security policy and convened a conference on education in 1955.

The combination of these developments, especially the dissemination of DeWitt's research, led Congress to increase pressure on the NSF to expand its educational work.[51] The funding for NSF summer teaching institutes rose dramatically through the mid-1950s. In January 1956, facing the normally unreceptive House Appropriations Committee, Waterman suggested the NSF's educational initiatives would aid in long-range manpower planning. Thomas had earlier been wary of Waterman's gestures to education—wary of the federal government's ability to usurp local control—but DeWitt's research on Soviet science had now convinced him otherwise. Thomas stunned Waterman by suggesting that, instead of the $3 million requested for high school teacher training, "if you could use $9 million or $10 million we are certainly prepared to give it to you." Phillips backed Thomas's offer by adding that "education of high school teachers should come ahead of everybody else."[52] The committee voted to increase the amount for NSF summer institutes from $3 million to $9.5 million, of which $9 million (equivalent to about $77 million in 2013) was strictly limited to secondary education.

While appreciating the windfall, Waterman and his colleagues feared it would set a troublesome precedent. In Waterman's subsequent Senate testimony, he asked for the financial amounts to remain the same but for the restrictive emphasis on secondary education to be eliminated. Waterman reminded the senators of the other federal agencies that had recently undertaken educational projects.[53] Waterman was caught in a bind: to keep Congress happy and increase the NSF's appropriation, he had to agree to substantially increase the NSF's educational work, but he fretted any such expansion would come at the expense of the more cherished research funding and graduate fellowships. Despite Waterman's entreaties, the Senate preserved most of the restrictions. The NSF, through no real desire of its own, substantially expanded its precollegiate work in order to secure increased funding from Congress.[54]

In July 1956, only a few months after the enlarged NSF budget for education in science had been approved, MIT's Jerrold Zacharias pitched a project to work on the physics curriculum to Waterman and the NSF assistant director of education, Harry Kelly. The NSF responded by creating the Physical Sciences Study Committee in the fall of 1956 with a $545,000 appropriation ($4.7 million in 2013 dollars). The physics committee initiated the Course Content Improvement Program of the NSF. Within a year, the NSF had begun planning future curricular work in biology and mathematics. Signaling this shift, and recognizing the new priorities of the Appropriations Committee, Waterman made high school education a central part of his presentation to Congress for the first time in early 1957.[55] Curriculum development and teacher training were no longer peripheral concerns of the NSF but central ways in which the foundation addressed doubts about the nation's scientific capabilities.

Sputnik's launch in October 1957 provided potent ammunition to those already pushing for reform. Scientists used Sputnik as an important emblem of the continuing need for military—and correspondingly scientific—financial support, while politicians who had long advocated curricular reform or federal education aid now believed Sputnik had proven them right.[56] Members of the American Council on Education, a consortium of representatives of higher education institutions, used their annual meeting the next week as a platform to connect Sputnik to the expansion of federal education initiatives. Before the end of October, Mary Ann Callan of the *Los Angeles Times* had begun a front-page, multipart series on American education and the looming crisis, linking it explicitly to the demand for scientists in this new space age.[57]

Initially responding coolly to the launch, President Eisenhower later appointed MIT's president James Killian as "missile czar" and formed the President's Scientific Advisory Committee. In turning to Killian—as well as to physicists Detlev Bronk and I. I. Rabi—for advice, Eisenhower firmly aligned himself with scientists who believed that the launch should be interpreted as an issue concerning training in the sciences generally rather than as one of military or technological priorities. A relatively small group of influential scientists drove the priorities of the advisory committee, the development of the curriculum projects of the NSF, and Eisenhower's public statements on science and education. The group's message was that education reform was critical to winning the Cold War.[58] The funding frenzy in the launch's aftermath provided plenty of evidence for what Eisenhower would later label the "military-industrial complex."[59]

Sputnik immediately raised the profile of the NSF and its education initiatives. Even after seven years of appropriations, the NSF had only 250 employees and a relatively meager $50 million budget for fiscal year 1958. By fiscal year 1959, the first NSF budget determined entirely after Sputnik, the annual NSF appropriation rose to $136 million, with nearly half marked for science education, broadly construed.[60] While the NSF already had more money than it wanted for primary and secondary school purposes prior to Sputnik, its Course Content funds increased at an even greater rate, from under $1 million to $6 million ($50 million in 2013 dollars). While spending in support of science increased slightly more than twofold over the two-year period, total spending on projects in education and curriculum development increased more than tenfold. When asked in early 1958 if the Appropriations Committee was adequately funding the NSF, Senator Hill responded that "I am sure you will find that the Appropriations Committee will be very sympathetic to anything needed for the National Science Foundation."[61] President Eisenhower, despite long-standing reluctance to involve the federal government in education, endorsed the Scientific Advisory Committee's recommendation that the national expenditure for education be doubled. Whereas in the early 1950s, Representative Phillips complained that the House wanted to appropriate money with an arithmetic progression, post-Sputnik everyone agreed the growth should be exponential.[62]

Supporting the work of scientists during the early Cold War played well politically; reformers in Congress on both sides of the aisle used the launch of Sputnik to push through far more extensive measures, including one of the first general aid bills for schools, the National Defense Education Act of 1958. The act aimed to improve education in science, mathematics, and foreign languages. Provisions included student loans, teacher training funds, graduate fellowships, state grants for counseling, testing, and guidance, as well as funding to upgrade teaching facilities, methods, and materials in these fields.[63] The coalition behind the bill cut across ideological divides: its major sponsors—Representative Carl Elliott and Senator Hill, both of Alabama—joined with liberals and other longtime advocates of increasing federal education expenditures. Most significantly, the coalition included a number of Cold Warriors hesitant to expand the government but willing to do so in the cause of national defense. Timing played a key role as well—in an election year, even an education bill somewhat tenuously linked to the national defense passed. After the National Defense Education Act's passage, Representative Carl Albert remarked that the bill will "meet the communist threat" by supporting the education "essential to the survival of democracy in the atomic age."[64] The United States had indeed committed to a prolonged Cold War of the classroom.

SMSG's Origins

In February 1958, a request for funding to start a new NSF program in mathematics—what would become SMSG—was granted in congressional hearings without comment or dissent. By this point, with the curriculum linked to concerns about manpower and lagging intellectual discipline, the creation of a new NSF effort in mathematics education was hardly newsworthy. The Physical Sciences Study Committee was already well under way, and the NSF had begun to diversify its curricular initiatives, sponsoring a conference with mathematician Mina Rees at Hunter College on the mathematics curriculum and supporting the biology textbook project of the National Academy of Sciences—National Research Council. Sputnik didn't create SMSG or the Course Content Improvement Program, but the 1957 launch ensured they would have the continued institutional and financial backing of the federal government.[65]

Only days after initial congressional funding was approved, NSF's assistant director of scientific personnel and education began to take specific actions to organize the mathematics reform effort. Although the official topic of the Conference on Research Training and Potential in Mathematics in February 1958 was the education of professional mathematicians, the conversation shifted rapidly to addressing the needs of the primary and secondary schools. A group of mathematicians, including Paul Rosenbloom, A. Adrian Albert, H. F. Bohnenblust, Marshall H. Stone, and Raymond Wilder, suggested that mathematicians should attempt to reform all levels of mathematics education.[66] Other mathematicians agreed with the proposal, and the attendees urged their professional association, the American Mathematical Society, to consult with other organizations to seek funds and move toward a solution. Conveniently, mathematicians were already scheduled to reconvene the next week for the general mathematics meeting of the NSF. At this gathering, which featured many of the profession's leaders, including Harvard's

Richard Brauer, president of the American Mathematical Society, and University of Kansas's G. Baley Price, president of the Mathematical Association of America, the assembled mathematicians moved to create SMSG as their contribution to the curriculum reform effort.

These academic mathematicians were willing to delve into pre-collegiate curriculum design not only because the NSF was generously funding such an effort but also because the physics group had shown how it might be done.[67] Of course, it helped that mathematician Edward Begle had already announced he was prepared to lead the operation. Begle possessed the requisite mathematical credentials as a student of Raymond Wilder's at the University of Michigan and of Solomon Lefschetz's at Princeton. After his appointment at Yale in 1942, Begle had begun to move away from research, writing an undergraduate calculus book and serving as secretary of the American Mathematical Society in the 1950s. By 1958, he was ready to try administration full time—he decided that he "wasn't good enough or pure enough" to continue research—and so took leave from his department at Yale to accept the SMSG job. Later Begle would also explain that he was particularly interested in the curriculum project because of the difficult experiences of his daughters in their own math classes.[68] He could have hardly expected that it would consume the rest of his professional life.

Begle arranged SMSG as a collection of semi-independent writing groups. Each summer, returning and new writers converged on a particular college campus to sketch chapters, revise drafts, and compose new textbooks. Begle believed it to be important to invite roughly equal numbers of mathematicians and teachers for each grade and create a collaborative atmosphere: mathematicians ensured the material was sound, while teachers ensured the presentation was sensible. (A Bell Labs mathematician, Henry Pollock, later recalled that mathematicians always had to run their ideas by a high school teacher to find out whether they would actually work in a classroom.)[69] Begle's insistence on the presence of actual teachers in writing teams—at roughly equal numbers with professional mathematicians—distinguished SMSG from other NSF programs. In the NSF biology and physics projects, for example, scientists ran the show entirely, attempting to produce instructional materials free from involvement with the educational establishment.[70]

For the first writing session in summer 1958, Begle created separate groups for each high school grade and one writing team for the middle grades. Henry Swain of New Trier Township High School in Winnetka,

FIGURE 2.1 Edward Begle, November 1961. School Mathematics Study Group Records, Dolph Briscoe Center for American History, University of Texas at Austin.

Illinois, chaired the ninth grade algebra group; Robert J. Walker, chair of mathematics at Cornell, the tenth grade geometry group; Frank B. Allen of Lyons Township High School in LaGrange, Illinois, grade eleven; and Donald E. Richmond, chair of math at Williams College, twelfth grade. For the combined seventh and eighth grade group he wanted short lesson plans instead of completed textbooks—while the middle grades needed the most improvement, Begle worried that "there is little agreement about what should be done." For this project, he appointed a group chaired by John Mayor of the American Association for the Advancement of Science. Begle established SMSG as a national organization by moving the writing sites around the nation and drawing on more than four hundred writers from over thirty-five states.[71]

By the end of the first month-long writing session in 1958, the writing teams had made impressive progress. The seventh and eighth grade group had produced thirteen units to test in the coming year, while each of the high school groups had an outline and the beginnings of a textbook.[72] These drafts would be revised and rewritten over the next three to four years with different summer writing teams, producing both preliminary and revised books for schools.

Despite its size, SMSG was effectively run by small committees, the members of which were responsible for appointing panels for projects, reviewing the progress of the group, and establishing SMSG's future direction. These few individuals made most of the critical decisions about SMSG and also served as the main spokespersons. That doesn't mean they specified content directly—following Begle's precept that any pedagogical idea was worth researching, SMSG writers could overrule Begle's own preferences for textbooks.[73] Nevertheless, the responsibility for the federal reform of mathematics education ultimately fell to a handful of mathematicians and teachers.

*   *   *

The new math emerged from the broad consensus that training in more rigorous, disciplined subjects would be good for the minds of students, and in turn would help the country win the Cold War of the classrooms. Congressional Appropriations Committees had steered the NSF into developing pedagogy and curriculum programs well before Sputnik entered orbit, and critics had been harping on the connection between mental discipline in the schools and the intellectual virtues of the populace for years. The NSF's programs were widely praised, in large part because they functioned to improve education through the intervention of scientists themselves. The NSF's programs avoided the appearance of federal interference with local curricular decisions and appeased those who wanted to reduce the influence of educational progressives while still supporting government-sponsored reform.

Only some disciplines, however, were understood to cultivate the desired sort of intellectual training. Illinois representative Noah Mason said in March 1958 that "in the good old days—30 or 40 years ago—our American schools used to teach and stress the 'Three R's.' We also insisted upon strict discipline. We were old-fashioned enough to believe that discipline was an important part, an essential part, a valuable part of school training." When prompted to answer whether everyone should learn mathematics—not just the talented few—National Academy of Sciences president Detlev Bronk replied that everyone needs the ability to think, which can be fostered by "an understanding of some fundamental and relatively few subjects," including mathematics. Approvingly citing the example of his father who took just two books with him for the 1849 California Gold Rush—Chaucer and Newton's *Principia*—Bronk explained that while his father was "no great scholar," he "had the satisfaction and the pleasure that was derived from the continued development of his mind which had begun under the rigorous system of education that was characteristic of that time." The hyperbole of such claims only served to reinforce the view that virtuous development of the mind arose from "rigorous" education in a few disciplines, particularly mathematics.[74]

Contemporaries portrayed SMSG as an exemplar of the NSF's attempts to spread the "rigorous" knowledge of disciplinary elites among American students. In 1962, for example, Benjamin DeMott portrayed SMSG in *The American Scholar* as fighting a battle against those who wished for university knowledge to remain isolated. While SMSG's effort started out a "shade utopian in character," the group has "created instruments with the aid of which many thousands of Americans will come, early in life, to a sense of the light and grace of a world once nastily bound in briars." The issue for DeMott was the ability of academics to speak publicly about the nature of their field, and he commended SMSG writers for doing so:

> They have demonstrated that the academic can serve the public interest brilliantly without meeching about the White House. [They] chose not to isolate themselves in tight graduate school hives; they were willing to expose their studies and their present modes of thought to the eyes and understandings of

humbler teachers and humbler students; they have not hesitated to accept responsibilities that other professionals too often cynically delegate to publishers' drummers or third-rate academic minds. Even if they had failed in the
task they undertook, they would for these reasons deserve praise and honor
rather than abuse. As it is, they have a claim to a tolerably high rank among
the exemplary intellectuals of this age.[75]

"Exemplary intellectuals of this age" were ones who brought knowledge
out of the "briars" of the university instead of delegating such educational work to "third-rate" minds. While it was fine to include educators, it was critically important that professional mathematicians led the
NSF's curricular projects.

The desire to promote science and mathematics was not primarily
about the technological or military utility of those fields. The NSF was
pushed into curriculum development in order to address concerns about
mental discipline. The emergence of the new math was evidence of the
success of professors' and politicians' attempts to link the development
of citizens' intellectual abilities to the national defense. As proponents
might have hoped, and critics certainly feared, the federal government
has never ceased using school legislation and funding to exert pressure
on schools, and politicians from both sides of the aisle have pushed for
educational reform through federal means.[76]

Discussion of textbooks' content was absent from the institutional
and political wrangling behind the creation of SMSG. Those pushing the
NSF to design new textbooks did not generally speak about the nature
of mathematical knowledge or any special relationship with intellectual
virtues. The critics and politicians supporting NSF's programs tended to
assume that students would be disciplined by mathematics—and disciplined in a desirable way. It was left to groups like SMSG to specify how
mathematics could and should cultivate mental habits. Their proposals
would not be what Congress expected.

CHAPTER THREE

The Textbook Subject

Mathematicians and the New Math

Created amid federal wrangling, SMSG was ultimately an effort spearheaded by professional mathematicians. The original eight members of the Advisory Committee were all academic mathematicians, and they determined the organization's structure and direction.[1] Well located in powerful positions within the ranks of their profession, these mathematicians began their curriculum work enjoying two advantages. First, the National Science Foundation and Congress had given them wide leeway to determine how to make the curriculum more rigorous. Second, mathematics and the sciences in general enjoyed great prestige for their role in helping to win World War II and their ongoing role in maintaining a Cold War military and technological lead on the Soviets.

SMSG's mathematicians took the charge to make the intellectual habits of American students more rigorous as an opportunity to introduce "modern" mathematics into the curriculum. They argued that developments over the first half of the century had fundamentally reformulated what it meant to do mathematics. Some SMSG writers explicitly argued mathematics was now best understood as the study of general abstract systems, divorced from potential physical applications. The curriculum project was their opportunity to inscribe this view of mathematics in millions of textbooks. Nevertheless, some of the earliest and most vehement critics of the curriculum were fellow mathematicians who argued that SMSG's textbooks misrepresented the true nature of mathematical knowledge. These critics believed recent developments in the field were ephemeral research fads, set apart from the main historical

trends of mathematics. New curricula rarely satisfy everyone, but critiques of SMSG's textbooks were particularly contentious.

This chapter uses the textbooks themselves to analyze how SMSG
presented the nature of mathematical knowledge. At least since the work
of historian and philosopher of science Thomas Kuhn, historians have
known that textbooks codify disciplines—not just by making facts and
prevailing dogma explicit but also by encoding procedures, norms, and
ways of thinking. Kuhn's treatment of textbooks was problematic. He
neither historicized the emergence of textbooks themselves nor considered the contingencies and ambiguities that accompany their use. Nevertheless, he correctly asserted that if historians wanted to know the nature
of a scientific discipline at a particular time, they ought to look at how
textbooks convey the facts, techniques, rules, and—most importantly—
exemplars that define the field.[2]

Secondary school science and mathematics textbooks also codify disciplinary practice and knowledge, but they differ in at least one crucial
respect: they are meant to be used even by those who will never become
practitioners. School textbooks mediate between academic fields and the
role those subjects play in general education. The primary and secondary school math classroom is peculiar, distinct from both the "everyday"
way people use mathematics and from the math done in universities.[3]
SMSG's writers claimed that previous mathematics textbooks had been
seriously deficient in their presentation of what mathematics was, what
mathematicians did, and what general benefits the study of mathematics offered. Nevertheless, the reform of textbooks at midcentury was not
primarily about accuracy, and there was nothing factually wrong with
the previous textbooks. (This was different from the situation in midcentury science education, where the textbooks were often considered
outdated.)[4] Mathematics was, therefore, a special case in the eyes of reformers: the books did not contain factual errors per se, but it was said
that they failed to fully exploit the benefits of learning math. All students should know how to add, yes, but all students should also know
about the structure and underlying nature of mathematical knowledge.

Research mathematicians, though, were not supposed to care about
school mathematics. American academic mathematicians had long *defined* themselves as removed from pedagogical concerns. That was why
professional mathematicians splintered into three organizations in the
first decades of the twentieth century. The American Mathematical Society was established for professional research, the Mathematical Asso-

ciation of America for the teaching of college-level mathematics, and the National Council of Teachers of Mathematics for high school mathematics instruction.[5] SMSG started as a collaboration of these societies, and membership on its initial steering committee was drawn from all of them. The unusual involvement of the American Mathematical Society in school mathematics was not unnoticed at the time, and some research mathematicians raised the concern that their colleagues shouldn't bother with precollegiate curricular issues. John Green, the society's secretary and a mathematician at UCLA, for example, complained to the organization's president (and Harvard professor) Richard Brauer that curriculum development is plainly "not Society business." Brauer's response was telling, suggesting not only that other efforts had failed to produce satisfactory curricula but also that the American Mathematical Society should never have abdicated responsibility for pedagogy in the first place and that it should now "accept the responsibility for clearing up the present mess."[6] SMSG's very existence testifies to the fact that mathematicians in this period believed they should play a substantial role in curriculum reform. *How* the textbooks should be reformed was forcefully contested, however. Mathematicians' approach to "clearing up the present mess" depended on their view of the underlying nature of mathematical knowledge.

New Images of Mathematics

SMSG's portrayal of the nature of mathematics was at least partially based on work that had already been done by the College Board's Commission on Mathematics, and Edward Begle explicitly encouraged his writers to take the commission's report as a "reasonable starting point."[7] Headed by Princeton mathematician Albert Tucker, the commission began work in 1955, and its report, published four years later, described the nature of mathematical knowledge by analogy.

> As a city grows, it becomes increasingly difficult to find adequate transportation from the center of the city to the outlying areas and the suburbs. The center is still the core of the city but the streets are too narrow and too congested for the newer sections to be reached as quickly as the needs of the residents require. For a time, systems of traffic lights and one-way streets suffice; but ultimately these patchwork methods, too, are found to be inade-

quate. Then there is constructed a limited-access freeway or expressway from the heart of the city to outlying points, bringing the newer regions effectively closer to the core.

The same thing was true about the changes in mathematical knowledge. "New and more efficient routes" in the foundations of the subject have been found which enable students to reach "modern mathematics" without "laboriously traversing all of the older content."[8] The claim was about both mathematics and pedagogy. Mathematics, like cities, had undergone structural reform, making it possible to navigate the whole more easily.

The analogy implied a degree of destructiveness. Midcentury cities did not keep all the old with the new but bulldozed some less useful structures in favor of new ones. Mathematics had to do this as well; subjects might be eliminated from areas of active research or teaching whether or not they had been proven "wrong." Certain topics might simply no longer be worth pursuing or might have been replaced by more powerful, direct, or efficient ones.

Without any acknowledgment, the report's authors had lifted this metaphor from an article by Nicolas Bourbaki. Appearing in English translation in *American Mathematical Monthly* in 1950, Bourbaki's article declared that it was time to reexamine the "internal life of mathematics" by noting that it was like "a big city, whose outlying districts and suburbs encroach incessantly, and in a somewhat chaotic manner, on the surrounding country, while the center is rebuilt from time to time, each time in accordance with a more clearly conceived plan and a more majestic order, tearing down the old sections with their labyrinths of alleys, and projecting towards the periphery new avenues, more direct, broader and more commodious." As did the College Board's report, Bourbaki's article claimed new routes should replace old labyrinths, enabling one to move from the center to the outlying regions most efficiently. Postwar mathematics and contemporaneous urban renewal shared an impetus: new structures justified some "tearing down" in order to form a "more majestic order."[9]

Borrowing the language of Bourbaki meant taking sides in the mathematical world. "Nicolas Bourbaki" was not a person but the pseudonym for a well-known and controversial group of mathematicians interested in completely reformulating the image of mathematics. With sensibilities forged by the intellectual milieu of the École normale supérieure in the

1920s, a number of talented young mathematicians including Henri Cartan, Claude Chevalley, Jean Delsarte, Jean Dieudonné, René de Possel, and André Weil formed Bourbaki the following decade. Despite, or perhaps because of, a culture of secrecy and exclusiveness (the name was both an inside joke from the École normale supérieure and a reference to a real nineteenth-century general in Napoleon III's army), the group's fame eventually extended well beyond its relatively modest initial goal of writing an analysis textbook in the "modern" style. Bourbaki's founders saw themselves as bearers of a new ideology of mathematics, and their project grew over the course of the 1930s into a series of volumes under the grandiose title *Éléments de mathématique*. Bourbaki's approach emphasized precise terminology, the modernization or elimination of "vague" concepts like spatial intuition, the importance of well-chosen axioms, and above all, the placement of structure at the center of mathematical practice. Even if their direct influence on the field may have often been overstated, Bourbaki stood for a new idea of what it meant to practice mathematics.[10]

Bourbaki's choice of a textbook for their central project indicates the importance of pedagogical concerns. Rather than changing the direction of research alone, Bourbaki's members pointed to the ways recent developments had fundamentally changed mathematicians' understanding of the nature of mathematical knowledge, necessitating a revolution in the way math was taught. The group held talks for high school teachers in Paris through the mid-1950s, arguing for a move away from traditional subject material in mathematics.[11]

Bourbaki's emphasis on the elimination of the "labyrinths" of outdated methods also influenced curriculum reformers. In a 1959 speech, one of Bourbaki's most outspoken members, Jean Dieudonné, proclaimed that "Euclid must go!" For Dieudonné, Euclidean geometry was "dead weight" with few interesting questions and little relevance. The useful bits might be profitably taught in a few hours, and the rest of the field had "as much relevance to what mathematicians (pure and applied) are doing today as magic squares or chess problems."[12] Mathematicians could find better ways to teach the rigorous—logical, analytical—methods of mathematics. Dieudonné speculated that only faith and tradition have kept Euclid alive in the curriculum. While focusing his attack on Euclidean geometry in particular, his conclusion concerned all outmoded subjects.

Dieudonné's speech provoked sharp controversy. He was speaking at

the Royaumont Conference, an international gathering concerning reform of the mathematics curriculum. Even among reformers, few were
willing to countenance the near-complete elimination of Euclidean geometry from the school curriculum. Nevertheless, Dieudonné's provoca-
tive claims were influential among the Americans in attendance: Tucker
of Princeton and the College Board, Howard Fehr of Teachers College,
and three members of SMSG—Begle, Robert Rourke, and Marshall
Stone. Stone had, in fact, delivered the opening address of the conference, encouraging precisely the sort of "modernization" of the curriculum that Dieudonné's views represented. One of SMSG's geometry writers later claimed that some of the group's work was a "partial answer"
to Dieudonné's "Euclid must go" speech. Looking back a decade later,
French mathematician René Thom suggested that it would be only natural for mathematicians "steeped in the ideas of Bourbaki" to write textbooks like SMSG's.[13]

SMSG and Mathematics as Structure

While Begle's appointment had been largely happenstance—he was
looking to move into administration full time just as the American
Mathematical Society needed someone to head the curriculum project—it resulted in an intellectual trajectory that rooted SMSG's conception of mathematical knowledge in Bourbaki-inspired soil. Perhaps
recognizing the divisiveness and hyperbole of Bourbaki's rhetoric, Begle never explicitly proselytized for Bourbaki's vision, but he was nonetheless a believer in their model of mathematics and echoed them when
he defined mathematics as a "set of *interrelated, abstract, symbolic systems*." Indeed, many of the American mathematicians involved were
closely linked to Bourbaki. Both Begle and Andrew Gleason, an SMSG-
allied reformer, had done mathematical research in a field heavily targeted by Bourbaki's reforms—topology—and Gleason later recalled that
he had likely been influenced by Bourbaki's emphasis on abstraction and
structure. Another key SMSG supporter, A. Adrian Albert, was the author of *Structure of Algebras*, a textbook centered on a favorite Bourbaki topic.[14] Although SMSG had semi-independent advisory boards,
and writing teams occasionally rejected Begle's advice, Begle and his
small group of similarly minded mathematicians retained a great deal of
power and were widely understood to control the direction of SMSG.

Of SMSG's central mathematician-writers, the most outspoken proponent of Bourbaki was Marshall Stone. Stone is widely credited with transforming the University of Chicago's mathematics department in the late 1940s, with the stated goal, as he later characterized it, of "elaborating a modernized curriculum."[15] One of his key hires in this respect was André Weil, a French mathematician closely involved with Bourbaki. Stone's own research into Boolean algebras focused on structures and classes rather than on the particular characteristics of elements themselves, a typical Bourbaki approach. Stone's research suggested the productivity of bringing the structure of mathematics to the fore, both in solving problems and in opening new areas of research, and he worked to bring similarly minded mathematicians to positions of prominence in the United States.[16]

In 1961, Stone published an article in *American Mathematical Monthly* explaining his view of the nature of mathematical knowledge and its implication for curricular reform. Stone argued that "modern" mathematics was truly "abstract" mathematics detached from physical applications. Mathematicians, Stone suggested, needed to "rethink" their "entire conception of education" since a "quiet revolution" had taken place separating mathematics from the physical world. It was now possible to "provide a nearly final answer to the question 'What is Mathematics?'" because "a modern mathematician would prefer the characterization of his subject as the study of general abstract systems, each one of which is an edifice built of specific abstract elements and structured by the presence of arbitrary but unambiguously specified relations among them." Stone proceeded to survey the various fields of mathematics and showed how a "fundamental unity" held. While not neglecting the "manipulative aspects" of the subject or the importance of those "who engage in making applications of mathematics," curriculum designers should push forward "abstraction and the discernment of patterns" as key elements of mathematical practice.[17]

These claims were inscribed into the textbooks SMSG produced. Examples could be drawn from any number of them, but the junior high textbooks were the most explicit in their engagement with the nature of mathematical knowledge. These school years traditionally fell between elementary arithmetic and secondary school geometry and algebra. SMSG positioned the junior high school textbooks as the "essential" bits of mathematics—what one needed to know about mathematics to "function effectively in our society."[18] Begle recognized at the organization's outset that this status made these volumes the most difficult to write, but perhaps also the most important.[19] The junior high school writing team wrote the actual textbooks in the summer of 1959 and finished the revised versions for the fall of 1961.

According to SMSG, whatever else they might have done, mathematicians certainly *did not* calculate. The junior high school series begins with a story of a mathematician telling his neighbor on a plane that he, in fact, did not do calculations with figures all day long—that was best done by a machine. Rather, "my job," the mathematician said, "is mainly logical reasoning." That meant "deductive reasoning," or reasoning from "if–then" statements.[20] The first problems students faced involved logic puzzles that required not counting or measuring, but careful reasoning about particular sets of information. Such problems attempted to convince students that math was more than an ever-expanding set of computational skills.

Moreover, mathematics was not a "completed" subject, analogous to Latin grammar, but rather an ongoing and expanding field of active research. "In the past 50 years," the SMSG junior high text proclaimed, "more mathematics has been discovered than in all the preceding thousands of years of man's existence."[21] Fittingly, the junior high curriculum ended with a chapter on unsolved problems, an attempt to dispel the idea that mathematics is a "dead and completed subject that was embalmed between the covers of a textbook sometime after Sir Isaac Newton." SMSG's writers did this in part by presenting then-unsolved (but relatively easy to grasp) puzzles, from Goldbach's conjecture to the question of how to pack spheres with the greatest density. The last few pages present David Hilbert's problems, a series of research queries that the eminent mathematician presented at the start of the twentieth century. The text noted that some have been solved but many remained open. "No one can say, of course," who will solve them, but "in all probability the problems mentioned in this chapter will not be solved by old graybearded professors . . . but by youngsters not much older than you." The eighth grade text rather bizarrely included a table listing the various subject classifications (from abelian groups to quasi-conformal complex functions) of the 1958 volume of the journal *Mathematical Reviews*.[22] In providing such a table—the terms of which would have likely been mysterious to undergraduates, let alone eighth graders—the authors tried to give a sense of the breadth of mathematics by overwhelming students with a massive list of unfamiliar words. The net effect of these efforts was to show students that mathematics was an open-ended mode of reasoning and that arithmetic calculations were only a minor topic.

SMSG also emphasized the contingency of mathematical symbols. Mathematics, SMSG suggested, had not always looked the same—the then-current set of symbols was simply the latest incarnation. The authors explained how numeration might have begun with "cave-men" and continued with the contributions of Egyptians, Babylonians, and Romans, each in turn maintaining different forms of mathematical representation. Symbols entailed choices, with one happening to be the adoption of the familiar "Hindu-Arabic" numerals. The point was not to trace the whole history of mathematical notation but to show that representations of mathematics were historically about human preference: "the decimal system is used in most of the world today because it is a better system."[23] Inscriptions were not mathematics, just the gradually improved tools of mathematicians.

Even the range of representations using the typical numerical symbols entailed choice. Thus, the text introduced different "bases" into junior high school classes. The usual numeral 15 was in base 10, representing one "ten" and five "ones," or $(1 \times 10^1) + (5 \times 10^0)$, just as 271 represented two "hundreds," seven "tens," and one "one," or $(2 \times 10^2) + (7 \times 10^1) + (1 \times 10^0)$. One could, of course, do it differently, and it would not have changed the *math* at all. Problems required students to write 45_{seven} (45, in base seven notation) in decimal notation:

$$45_{seven} = (4 \times 7^1) + (5 \times 7^0) = (3 \times 10^1) + (3 \times 10^0) = 33_{ten}$$

Exercises included multiplication and division in various bases, conversion between bases, rules for divisibility in other bases, etc.[24] The authors intended to deemphasize the most familiar representation of *doing* mathematics—the written forms of elementary arithmetic. The essence of mathematics was reasoning, not ciphering.

Even within the realm of mathematical calculations, the usual arithmetic operations constituted only a small minority of possible computations. The authors introduced modular arithmetic as a "new kind" of calculation, that is, an operation acting on a finite set of elements. Clocks are often used to introduce some of the peculiarities of modular arithmetic, because hour hands can only point to a finite set of numbers, rang-

3. Here is a table for addition (mod 5).

+	0	1	2	3	4
0	0	1	2	3	4
1	1	2	3	4	0
2	2	3	4	0	1
3	3	4	0	1	2
4	4	0	1	2	3

Copy and complete the following table.

(Mod 5)

b	a	additive inverse of a	b - a	b + additive inverse of a
0	1	4	0 - 1 ≡ 4	0 + 4 ≡ 4
2	1		2 - 1 ≡	2 + 4 ≡
4	1			
1	2			
2	2			
3	2			
2	4			
3	4			
4	4			

FIGURE 3.1. An exercise in modular arithmetic from SMSG, *Mathematics for the Junior High School: Student's Text*, 1:557.

ing from one to twelve. So, in the "clock arithmetic" version of modular arithmetic, $4 + 4 = 8$ but $4 + 9 = 1$ (four hours after four o'clock would be eight o'clock, but nine hours after would be one o'clock). Students were asked to recognize how some familiar concepts of regular arithmetic—like "additive inverse" in the example above—could carry over to these new operations.

The table in Figure 3.1 provided students the basic arithmetic facts directly; the point was not to calculate but to recognize the similarities and differences between the structure of this new form of arithmetic and that of the usual arithmetic calculations.

SMSG presented modular arithmetic as an alternative mathematical "system," or "a set of elements together with one or more binary operations defined on the set."[25] It exemplified a mathematical system with its own elements, operations, and properties, just like the common arithmetic students had already learned. After suggesting that usual arithmetic was only one of many possible operations and the numbers 1, 2, 3, . . . just one particular set of possible symbolic representations, the authors' final surprise was to assert that the objects of mathematical calculation need not even be numbers. The possible spatial manipulations of a playing card, for instance, constituted a mathematical system because one can think of the various orientations of the card as the elements and the procedures of flipping and rotating as operations transforming one element (orientation) into another. Again, properties and concepts from elementary arithmetic like "inverse" and "commutativity" could be defined. Mathematical reasoning involved defining and exploring the properties of systems, not learning the facts of one particular system of calculation.

The idea of mathematical systems did more than expand students' ideas about what "counted" as mathematics. The authors also wanted to show how mathematicians made their choices. Mathematicians, SMSG claimed, proceeded by elucidating and analyzing the structure of mathematical systems, rather than by blindly following rules or solving a collection of seemingly indistinguishable problems. By introducing un-

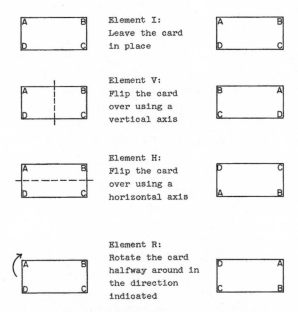

FIGURE 3.2. Calculation need not involve numbers at all; a mathematical system might be constituted from the various orientations of an index card. SMSG, *Mathematics for the Junior High School: Student's Text*, 1:566.

familiar forms of mathematical practice, the textbooks aimed to teach students to recognize that it was a way of thought which defined mathematics, not a fixed set of techniques or objects.

The basis for many of the systems in SMSG's textbooks was the usual arithmetic with positive integers. Mastery of arithmetical properties would enable students to give coherent accounts of other forms of calculation. For example, if a mathematician wanted to operate with negative integers, SMSG implied that he would first define the relevant procedures by considering the properties of multiplication with positive integers. If positive integer multiplication was commutative, distributive, and associative, say, then when introducing negative integers, one ought to pick a definition for multiplication such that those properties still held. The authors gave students this argument for why two times negative three (2×-3) should be defined as equal to negative six (-6):

$$3 + (-3) = 0$$
$$2 \times (3 + (-3)) = 2 \times 0 = 0$$
$$2 \times (3 + (-3)) = (2 \times 3) + (2 \times (-3)) = 6 + (2 \times (-3))$$
$$6 + (2 \times (-3)) = 0$$

Namely, students start with the knowledge that -3 is the representation of the number which when added to 3 yields 0.[26] By having assumed that the distributive property holds, the authors show that $2 \times (-3)$ and 6 must also be additive inverses, and since there is only one additive inverse for a given number, then $2 \times (-3)$ symbolizes the same number as -6.

Similarly, they suggested that $(-5) \times (-4) = 20$ by noting

$$4 + (-4) = 0$$
$$(-5) \times (4 + (-4)) = (-5) \times 0 = 0$$
$$(-5) \times (4 + (-4)) = ((-5) \times 4) + ((-5) \times (-4)) = 0$$

Students already knew from the ability to multiply a negative and a positive that $(-5) \times 4 = -20$ and the above reasoning suggested that $-20 + ((-5) \times (-4)) = 0$. By concluding that -20 and $((-5) \times (-4))$ are additive inverses, it follows that $(-5) \times (-4) = 20$.

Obviously, the purpose of these demonstrations was *not* to train students to compute quickly or to "rigorously" prove every assertion (at this juncture they omitted, for example, to show students that there is only one additive inverse for a given positive integer). The text could have simply started with the *rule* that 2 × (−3) yields the same result as taking the negative of the known calculation 2 × 3. But, the text suggests, students should not be compelled to believe such unmotivated rules. By emphasizing agency in choosing and asserting certain properties, one preserves the idea of mathematics as an open-ended practice. Students still needed to learn the usual arithmetic facts, but not as ends in themselves. "Understanding" here meant more than simply knowing the basis for algorithms—it meant recognizing mathematics as a practice, as systematic reasoning generally.

SMSG's books represented a rejection of mathematics as either a fixed set of skills for practical application or as a list of facts to be memorized. In doing so, the group's writers were following the lead of Stone and many others in deemphasizing math as primarily rote computation. Detlev Bronk, president of the National Academy of Sciences in the 1950s and then chair of the governing National Science Board of the NSF, complained around the time of SMSG's founding that in most mathematics books "there is too much emphasis upon facts, and too little emphasis on training in the ability to think." For SMSG's allies, structure, not technique, characterized the discipline. The NSF later elaborated what methodological role a new mathematics curriculum might play when it explained that SMSG would be formed to make mathematics "a way of thinking rather than a system of artificial devices to solve problems." Math was not widely useful because of some particular set of facts or processes but because it was itself a way of thought. Stone himself said as much in a 1957 speech: "science is reasoning; reasoning is mathematics; and therefore, science IS mathematics."[27] "Modern" math was a model of rigorous thought, so one needed only to introduce students to the way in which mathematicians think.

Stone and others involved with SMSG condemned the claimed persistence of textbooks that facilitated the study of mathematics as nothing more than endless facts and drills. While it is certainly possible to find textbooks of this sort in the period, as a general matter, this claim elided the differences between textbooks and classroom practices: no matter what textbook is in use, a teacher might rely upon exercises and activities which suggest that mathematics is about mental gymnastics and memorization more so than problem solving or creativity.

Complaints like Bronk's also drew upon the long-standing idea that math class should train students to think. As early as debates about whether to use Nicholas Pike's *Arithmetic* (1788) or Warren Colburn's

First Lessons in Arithmetic (1825), American math textbooks were evaluated on the basis of how they cultivated mental discipline generally.
While Pike emphasized the importance of memorizing arithmetic rules
and then applying them to various examples, Colburn's innovation was
to reverse rule and example: instead of presenting rules, he presented
simple examples in an effort to lead children to form rules for themselves.[28] Contemporaries understood the differences between the textbooks to be about differences in reasoning. The *American Journal of
Education* proclaimed in 1827 that rule-based methods failed because
a student would not have "been called upon, in this process, to exercise
any *discrimination, judgement, or reasoning. . . .* So that, as a discipline
to his mind, he has derived none, or very little advantage."[29] Conversely,
a reviewer in the journal for the Calvinist Princeton Theological Semi-
nary claimed that inductive methods would inhibit students' ability to
incorporate future knowledge and would ultimately undermine authority by erasing the traditional grounding of rigorous knowledge in rules.
Other reviewers addressed parents who wished to have arithmetic taught
"the good old fashioned way."[30] Even—perhaps especially—at the most
elementary levels, evaluating mathematical methods entailed assumptions about the virtues of intellectual training.

The "usefulness" of math as a mode of thought was still emphasized
in the early twentieth century, but on almost diametrically opposed
grounds. The 1920s and 1930s featured the "civic phase" of mathematics
education, characterized by the rise of "functional" and "general" mathematics.[31] By the Great Depression, math books overwhelmingly emphasized the social value of learning mathematics, rather than its importance as an intellectual discipline or its role in scientific inquiry. The
epitome of this trend was the Progressive Education Association's *Mathematics in General Education* (1940). The Progressive Education Association noted that more students had been enrolled in school with no plans
for college, students for whom traditional mathematics courses made little sense. Increasingly, the authors claimed, mathematics ought to be in
the service of education focused on the "wholeness of the personality of
the learner and the wholeness to which he responds." Students rejected
mathematics because of its inappropriateness for contemporary life—the
subject seemingly did not help students respond and adapt to their envi-
ronment. Mathematics might still be made useful, however: "The Committee shares the conviction that, *divested of much of its conventional*

content and formal organization, mathematics as a mode of thought and an instrument of analysis has an indispensable function in general education."[32] The association suggested mathematics courses that featured eminently practical topics such as measuring variation in student growth or comparing bank accounts. By stripping away the content and formal organization, teachers could make mathematics worthwhile for modern students.

If *Mathematics in General Education* was an extreme case, the claim that mathematics was applicable far beyond its nominal techniques had many manifestations and plenty of supporters in the twentieth century. A well-known (if little followed) 1923 Mathematical Association of America report promoted "functional thinking" generally through an understanding of mathematical functions. During World War II, the textbook *Mathematics for Victory* emphasized the numerous applications of the subject in the war effort. And after the war, Stanford mathematician George Pólya claimed that mathematics provided students a general set of problem-solving heuristics.[33] Bronk was undoubtedly justified in his belief that some textbooks overemphasized facts, but math had long been understood to have the potential to cultivate "ways of thinking."

For SMSG, authors from Pike to Pólya may have been correct in asserting a role for math as a mode of thought, but they did not place enough emphasis on the field's fundamental nature. SMSG's authors, following Stone's lead, portrayed mathematics as the study of structures and systems. The practical uses *and* the pedagogical justification of mathematics emerged from a proper appreciation of its foundations.

Algebraic and Geometric Reasoning in High School Textbooks

SMSG produced textbooks for a variety of high school subjects, but the algebra and geometry textbooks were by far the most influential and best encapsulate the larger pedagogical aims of the organization.[34] Traditionally, the analytic methods of algebra allowed students to manipulate symbols mechanically or experimentally to arrive at new relationships between quantities, while geometry emphasized using primitive postulates to systematically construct propositions about spatial figures. Although geometric and algebraic techniques have been intertwined in mathematical research at least since the seventeenth century, the distinction remained pedagogically important because the two subjects entailed different means of arriving at reliable knowledge. By the time SMSG released algebra and geometry textbooks in the early 1960s, however, its writers wanted to emphasize the *unity* of mathematics, not its variety. The presentation of math as a general exploration of systems and structures implied that the differences between algebra and geometry were no longer important. The geometry text, the teachers' commentary announced, introduced algebraic concepts and techniques early on, because "a great deal of the traditional material of geometry *was really algebraic all along*." The book's postulates freely mixed concepts taken from both subjects, for example, in the correspondence between a geometric line and the real numbers. SMSG's presentation implied a fundamental shift away from distinguishing geometry from algebra. Geometry was just one more form of mathematical reasoning, not a privileged method of revealing spatial truths.[35]

The practices of geometry and algebra inculcated the *same* intellectual discipline in students. The novelty of the new secondary textbooks was not in their content: SMSG still covered the well-worn conclusions of Euclid and the typical techniques in algebra instead of, say, projective geometry or abstract algebra. Rather, the authors believed that the theorems of Euclidean geometry and the techniques of algebra were simply equivalent ways to demonstrate for students the unified and systematic nature of mathematical knowledge. Both subjects instantiated the discipline of mathematical reasoning.

Geometry

The method by which SMSG intertwined algebra and geometry is exemplified by the group's treatment of the Pythagorean theorem. Euclid's *Elements* presented a lengthy proof based on the construction of squares and parallelograms on the sides of a right triangle. By comparing angles and lengths, Euclidean geometry textbooks reached the familiar conclusion that the length of the hypotenuse squared is the same as the sum of the squared lengths of the other two legs. Euclid's text emphasized constructions and comparisons; SMSG instead presented a proof based in symbolic analysis. After briefly mentioning the computation of the area of triangles and parallelograms, SMSG's *Geometry* directed students to consider a square of side $a + b$ formed by aligning four triangles of sides a, b, and c:

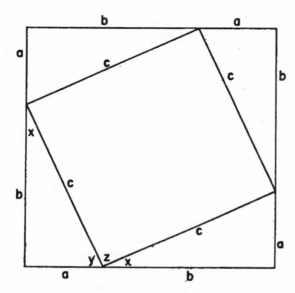

FIGURE 3.3. An algebraic-geometric derivation of the Pythagorean theorem. SMSG, *Mathematics for High School: Geometry*, 339.

The authors then asserted that this interior rhombus was a square by showing that it had right angles (they noted that angle y plus angle x yielded a right angle, meaning angle z was also a right angle). Then, the area of the large square was the same as the area of the smaller square plus the area of the four triangles. Symbolically they equated areas $(a + b)^2 = a^2 + 2ab + b^2 = c^2 + 4(\frac{1}{2})\, ab$, and then algebraically manipulated the equation to show that $a^2 + b^2 + 2ab = c^2 + 2ab$. The Pythagorean theorem, $a^2 + b^2 = c^2$, followed by subtracting $2ab$ from each side.[36]

The treatment of the Pythagorean theorem was indicative of SMSG's general approach to geometry because it used the properties of algebraic manipulation within geometric demonstrations. It was not a matter of conveying new facts to students; the practice of moving from postulates to conclusions was the important thing.[37] Geometry was not about the constructability of figures and the elucidation of their properties but rather exemplified systematic reasoning using postulates and theorems.

Such pedagogical choices were predicated on broader disciplinary developments, especially those concerning Euclidean geometry. While Euclid's theorems had long been synonymous with incontrovertible spatial truths, the proliferation of non-Euclidean geometries in the nineteenth century challenged that association. Mathematicians began to explore alternative geometric postulates and discovered that certain "evident" assumptions—such as the division of a plane into two distinct regions by a circle—were quite difficult to prove.[38] Increasingly, complex or so-called monstrous figures prompted mathematicians to expand the scope of allowable conceptual entities. By the end of the century, the relation of demonstration to understanding had been upended: the traditional order of things—objects exist in the world, one can describe them us-ing mathematical concepts, set out self-evident postulates about those objects, and then reason mathematically—was reversed. Now one began with postulates, developed their implications, and only then examined what—if anything—they might say about the world.[39]

By 1900, it was possible to say that the objects of geometry had little or nothing to do with the empirical spatial world, and that geometric con-clusions were no more reliable than those of other areas of mathematics. Thus, American mathematician Oswald Veblen had to specify in 1904—because it could no longer be assumed—that his geometric axioms "cod-ified in a definite way our spatial judgments." Slightly later, David Hil-bert openly questioned geometry's connection with spatial experience by noting that while "two points define a line" is a geometric statement, our terms "points," "lines," and "planes," could just as easily be re-placed by "tables," "chairs," and "beer mugs." The meaning of a "line" need not refer to anything spatial at all. By the 1920s and 1930s, Har-vard mathematician George D. Birkhoff coauthored a textbook, formed from his own axioms, which emphasized that geometry was about teach-ing general reasoning, not presenting a style of argument inherited from ancient Greece. Rather, the "prime concern" was simply "to make the students articulate about the sort of thing that hitherto [they have] been doing quite unconsciously."[40] The achievements of Veblen, Hilbert, and Birkhoff laid the intellectual groundwork for SMSG's *Geometry* to draw freely from algebra and other disciplines. It provided simply one more example of—and training in—structured mathematical knowledge, not a list of spatial facts or an ossified form of reasoning.

Algebra

If geometry had long been an exemplar of systematic reasoning, algebra had not. From early Arabic treatises on restoring and balancing equations to the later introduction of variables and signs for operations, algebra was historically an art of manipulating symbols. The algebraic practices involved in solving equations, simplifying expressions, and exploring the properties of numbers were highly formalized and often explicitly mechanical. SMSG's textbooks instead attempted to ground algebra, like geometry, in structured demonstrations, and thereby avoided having algebra simply "[degenerate] into just so much formalism." Its writers did this in two ways. First, they suggested that the techniques of algebra were only useful insofar as they revealed the structure of numbers. Rather than thinking of the techniques as meaningless symbolic games, algebraic simplification allowed students to understand the underlying relations between numbers, and therefore the structure of the number system. "The problem [of factoring]," the text informed students, "is to write a given polynomial, which we consider to be of a certain type, as an indicated product of polynomials of the *same* type." The authors emphasized the importance of algebraic types, that "for each value of its variables a rational expression is a real number."[41] Second, general conclusions about numbers required mastery of algebra. Specific facts about expressions involving properties of rational and real numbers could not be learned only by calculation—there were simply too many numbers. They required starting with postulates and moving to general conclusions, not conclusions about one particular number or figure.

Proofs consequently entered the introductory algebra textbook to show students how algebra allowed them to know something about numbers in general. After having learned elementary arithmetic, for example, students were supposed to know that "two numbers whose sum is o are related in a very special way," namely, that if one adds a particular number to 4 and gets o, then the number is an additive inverse of four. Then, students were asked whether there can be more than one additive inverse for a given number. "All our experience with numbers tells us 'No, there is no other such number.' But how can we be absolutely sure?" The authors suggested that this could be settled algebraically using the underlying properties of the system:

Suppose z is any additive inverse of x, that is, any number such that $x + z = 0$ . . . We use the addition property of equality to write

$$(-x) + (x + z) = (-x) + 0$$

We have then that

$$((-x) + x) + z = (-x)$$

$$\cdots$$

Then

$$0 + z = (-x)$$

And finally

$$z = (-x)$$

Algebra enabled students to be absolutely sure that every number has only one additive inverse because the symbols can be manipulated algebraically to conclude that given some x, the only additive inverse is $-x$, which is the opposite of x. Like the geometry text, the proof began with postulates and elements, and reasoned to a conclusion based on those postulates. Students were told as much: "What we have done is to use facts which we have *previously* known about all real numbers in order to argue out a *new* fact about all real numbers, a new fact which you certainly expected to be true, but which nevertheless took this kind of checking."[42] Mathematics was a structured process of moving from intuition to reliable conclusions.

Such proofs were at the core of what SMSG wanted to present as the goal of algebraic knowledge: the ability to use symbols to know things *in general* about numbers, in exact analogy to the way geometric proofs would enable students to know *in general* about triangles or circles. Teachers were repeatedly reminded that algebraic demonstrations risked devolving into "mechanical manipulation of symbols." Instead, teachers ought to emphasize that the techniques of algebra were a consequence of the properties of the number system. That is, they were yet another demonstration of the systematic nature of mathematical knowledge and practice.[43]

SMSG's algebra and geometry textbooks were not new in the sense of including new topics. The objects themselves did not much matter, so the traditional topics could serve just as well as newer ones. Rather, the novelty of the books was to recast the standard topics of school mathematics as opportunities for students to learn about the importance of mathematical structures. As the College Board's 1959 report had insisted, each of these fields should be "studied or taught with an eye to the fundamental nature of the subject," and this "more general point of view tends to make the whole subject more understandable." SMSG wanted teachers to emphasize that the textbooks' goal was to enable students to "achieve some appreciation of the nature of mathematical systems."[44] The books exemplified Begle's—and Stone's and Bourbaki's—conception of mathematics as a "set of interrelated, abstract, symbolic systems."

Mathematician Critics

SMSG's ideas about the nature of mathematics came under scrutiny from mathematicians almost from the start. As early as the fall of 1961, just as SMSG released its first set of revised textbooks, Harvard's George Carrier described a "bitter controversy" that had arisen over secondary school education within the mathematics community.[45] What was taught in the high schools suddenly mattered a great deal to mathematicians. Partly this was a result of SMSG's prominence and import. Unlike the products of any other math curriculum effort, SMSG's texts had the backing of the American Mathematical Society and so took on the appearance of a definitive text. This aspect was particularly problematic because SMSG had posited the "actual" underlying nature of mathematics as fundamental to its presentation. Its critics challenged the curriculum on both ontological and pedagogical levels: that mathematics was actually just so and that students ought to learn math in that way.

Carrier's characterization was made as part of a much broader expression of concern over the new curriculum on the part of applied mathematicians. The Society for Industrial and Applied Mathematics convened a discussion in November 1961 on precisely the issue of how mathematics should be taught. The panel included Carrier, H. J. Greenberg of IBM's Watson Research Center, Richard Courant, the head of New York University's applied mathematics institute, Paul Rosenbloom, a math professor at the University of Minnesota, and C. N. Yang, a Nobel Prize winner for his work on particle and statistical physics. Courant began by lamenting Stone's article in the *American Mathematical Monthly*. For Courant, Stone's vision of mathematics divorceable from physical reality was a "danger signal." The "fashion" of abstraction was not simply "half" true but mischaracterized the actual practice of mathematics. It might be easier to just give up the dream of math as a description of "substantive reality," but Courant warned that "the life blood of our science rises through its roots," and "these roots reach down in endless ramification deep into what might be called reality." Therefore math "must not be allowed to split and to diverge towards a 'pure' and an 'applied' variety." Much of the responsibility should be shouldered by teachers of math, who need to avoid "uninspiring abstraction," "isolation of mathematics," and "catechetic dogmatism," in favor of emphasizing a "close interconnection between mathematics, mechanics, physics, and other sciences."[46]

While Greenberg, Yang, and Carrier were broadly supportive of Courant's views, Rosenbloom dissented by suggesting that mathematics' applicative power came from abstraction and structure, not from attention to preexisting external problems. An SMSG author himself, Rosenbloom quoted physicist Paul Dirac's complaint that physical reality can be a "straitjacket" for a mathematician's imagination. A mathematician, Rosenbloom suggested, "finds it much more fruitful to create a mathematical model and then look around for a physical interpretation." Rosenbloom explained how he had initially organized the junior high school SMSG books around the physical applications of the material, but then asked a theoretical physicist, Francis Friedman, to examine them. Friedman told Rosenbloom that the "most practical thing in this book is the chapter on the finite mathematical systems because this shows students how to create new models."[47] Rosenbloom's comments echoed the way SMSG itself explained mathematics' usefulness. "One of the most important activities of modern mathematicians," SMSG's writers told teachers, "is the search for common attributes or properties often found in apparently diverse situations or systems. . . . Frequently, the systems developed out of the intellectual curiosity of mathematicians and their search for patterns in diverse abstract situations have been exactly the tools needed and seized upon by scientists in their attack on the problems of the physical world."[48] Acting out of curiosity, in the abstract and intrinsic "interest" of uncovering properties and structure, mathematicians had found tools amenable to the study of the physical world. The utility of mathematics, according to Rosenbloom and SMSG, emerged through the way students learned to think structurally and systematically.

The debate was not about whether math was useful but about the relationship between the subject's practical applications and its underlying nature. World War II had quite spectacularly demonstrated the ways seemingly esoteric mathematical sciences could be broadly applied. Because few disciplinary boundaries were respected during the mobilization, however, postwar institutional arrangements for the burgeoning field of applied mathematicians were largely ad hoc. (In countries where the war mobilization of scientists and mathematicians had been less aggressive, the discipline was even less developed.) SMSG's approach apparently caused great concern for applied mathematicians trying to establish the relevance of their new field. The Society for Industrial and Applied Mathematics itself had only been founded in 1951, formal "applied mathematics" departments were relatively new in the 1960s, and the conception of an administratively and intellectually separate field of "applied mathematics" was still precarious. Over a decade after the end of the war, MIT's H. P. Greenspan was still addressing the Mathematical Association of America to explain what "Applied Mathematics as a Science" actually meant. Noting the recent initiatives of institutions to create separate programs of applied mathematics, Greenspan attempted to define the field and dispel some of the rumors about it, particularly that applied mathematicians only performed a "technical service." The chair of the conference on education in 1961, H. J. Greenberg, had similarly opened the discussion by lamenting that applied mathematics was quickly becoming "something of a stepchild" or even an "out-of-step child," "looked upon with equal disinterest by mathematicians, physicists, and engineers." Applied mathematicians were clearly on the defensive in this period despite the success of wartime mathematics, as the American Mathematical Society rejected proposals to form special divisions for applied mathematics, while Fields Medal committees and international conferences increasingly eschewed "applied" work. SMSG's curriculum provided a venue for those like Greenspan, who desired to correct the "overemphasis" on "pure mathematics" and reassert the importance of applied mathematics within the school curriculum.[49] Disagreements about SMSG's textbooks pointed directly to divisions within mathematics itself.

SMSG's critics did not solely focus on the group's treatment of mathematics' applications. They also complained that SMSG had misgauged the relationship between mathematical practice and pedagogy. Foremost on this point was one of Courant's colleagues, Morris Kline. Kline was suspicious about SMSG's textbooks, and he published one of the first critiques of the new math in an October 1961 article. Under the heading "Unqualified Indictment," Kline concluded that "the direction of reform has been almost wholly misguided." Accusing the reformers of desiring change for change's sake, rather than basing their work on evidence of the failures of past instruction, Kline rejected the "old" version of math as a stultifying set of facts and rules but did not want to replace it with mathematics that emphasized formal structure above all else. While differences existed among the various new textbooks, he noted that "in the main they have a common curriculum" which focuses on "relearning in great logical detail the properties of numbers," wasting time that "would be better utilized in taking the students on to more advanced areas." The reforms, he warned, would do more harm than good, because "all of the desirable changes amount to no more than a minor modification of the existing curriculum, and all talk about modern society's requiring a totally new kind of mathematics is sheer nonsense."[50]

Most mathematicians first came across Kline's critique in a letter signed by dozens of mathematicians and published in 1962. Even though no author was listed, it was an open secret that Kline led the effort, alongside Pólya, Lipman Bers, and Max Schiffer. The letter put forward a set of principles for new curricula, three of which dealt specifically with the relationship between the nature of mathematics and instruction in the subject. First, the letter postulated that "knowing is doing" and that "knowing how" should be emphasized in the classroom. Students who understand concepts but cannot calculate or manipulate symbols, the authors claimed, do not know mathematics. Math's relevance for students emerges from concrete situations, not formalisms. Second, concepts should be introduced by examination of cases and then connected to scientific applications of mathematics. The significance of mathematics lay in the extent and success of its applications, not in its internal structure or consistency. Third, the "traditional" subjects—algebra, geometry, arithmetic, etc.—which were "fundamental" fifty or a hundred years ago were still foundational. The problem with the traditional curriculum was not an overemphasis on these subjects, but rather that they were taught in isolation from other domains of knowledge, especially the physical sciences, and thus came across as isolated tricks or rote manipulations. Coherence came from context, not from within. Noting that they could not "enter here into detailed analysis of the proposed new curricula," the authors nonetheless concluded that there were certain aspects "with which we cannot agree." The letter was not critical of any specific curriculum, but few were fooled; when the *New York Times* later covered the controversy, its reporter correctly noted that the letter was read as growing out of Kline's criticism of SMSG.[51]

SMSG's leader published his own response a few months later. Begle deflected the criticism by noting that nearly all the principles were also espoused by SMSG writers. (What did not need to be said was that the sites of this exchange—the *American Mathematical Monthly* and the *Mathematics Teacher*—were the organs of the Mathematical Association of America and the National Council of Teachers of Mathematics, two of SMSG's sponsors.) In subsequent inquiries, Begle found few of the mathematicians who signed the letter to have any knowledge of specific curricular materials. Those who were familiar with specific reforms often either believed the letter to apply only to non-SMSG materials or signed the letter as an ironic gesture, indicating their belief that its principles were in fact consistent with SMSG's work.[52] Begle ultimately condemned the letter as unhelpful in that such comments without corresponding reform efforts would never have any positive impact. Begle's response emphasized that the criticism should not be taken to imply that some large group of professional mathematicians rejected SMSG's textbooks, but the whole episode nevertheless revealed fissures within the professional community. One of the authors of the letter, Lipman Bers, had been worried for years that SMSG might appear as an "artificially united front" that minimized very real differences of opinion among mathematicians.[53] Even if relatively few mathematicians were as engaged with curriculum as Begle, Stone, or Kline, it was widely understood to be consequential for the profession as a whole.

The letter emphasized two additional pedagogical claims that Kline had long been pushing and which cut to the heart of the controversy. The first was the assertion that pedagogy ought to follow the "genetic" principle. Mathematicians may care about structure and logic, but students wouldn't. Instead, Kline proposed organizing the curriculum genealogically. The 1961 letter described this by claiming, "the best way to guide the mental development of the individual is to let him retrace the mental development of the race—retrace its great lines, of course, and not the thousand errors of detail." As Thom, the French mathematician critical of Bourbaki and SMSG later clarified, "Pedagogy must strive to recreate (according to [Ernst] Haeckel's law of recapitulation—ontogenesis recapitulates phylogenesis) the fundamental experiences which, from the dawn of historic time, have given rise to mathematical entities." The role of history here was explicit, and Kline's endorsement of the "genetic" principle was derived in part from his extensive historical research. Kline's surveys of the history of mathematics laid the foundation for the claim that historical order mattered—presenting logical structures as the actual nature of mathematics did little justice to centuries of mathematical development.[54]

The "genetic" view was the only principle expressed in the letter that Begle rejected entirely. He countered by explaining the absurdity of forcing students to compute with all the historical number systems before learning the "far more efficient" system of decimals. Similarly, Stone complained that "reliving the experiences of the race" was an "old chestnut" and an incoherent precept given the divergent intellectual histories of mathematics around the world, even in the West.[55] For Begle, Stone, and other SMSG partisans, mathematics in textbooks should resemble mathematics in contemporary practice, and this practice was ultimately about investigating structure. "Modern" mathematics had obviated the need to waste time tracing historical developments when an understanding of the underlying structure of mathematics provided students the essential reasoning skills.

The second pedagogical critique drew on Kline's long-standing opposition to the claim that mathematical structures were indeed at the heart of mathematical practice. He charged that the newer curricula did not communicate the nature of mathematics as a "living, vital, and highly significant subject," with the result that the curricula distorted the "primary value of mathematics," which was as "the language and essential instrument of science." Kline had done his doctoral work in topology, but after he joined New York University in the 1930s, Courant convinced him that "the greatest contribution mathematicians had made and should continue to make was to help man understand the world about him."[56] For Kline, groups like Bourbaki—far from their stated goal of providing an entirely new way to conceptualize the discipline— were a passing fad. An emphasis on logical, abstract structures misconstrued the history of mathematics and the way most mathematicians actually go about their work.

The interconnection of the history and nature of mathematics came together in the image of mathematical development that Kline supplied to rival Bourbaki's metaphor of urban renewal. He repeatedly pointed to how math is like a "great tree ever putting forth new branches and new leaves but nevertheless having the same firm trunk of established knowl-

edge. The trunk is essential to the support of the life of the entire tree."
Such images were in fact far more common than those portraying math's
development as destructive. (Kline himself claimed he took the metaphor from the work of the late nineteenth-century mathematician Felix
Klein.) Just prior to the new math controversy, Glenn James, the man-
aging editor of *Mathematics Magazine* and a professor at UCLA, had
published a book titled *The Tree of Mathematics*. The frontispiece and
cover displayed a tree with roots labeled "algebra" and "plane geometry"; a trunk of "calculus"; and branches including "topology," "calculus
of variations," and "probability." This common organic metaphor emphasized both the permanence and the cumulativeness of mathematical
knowledge as well as the dependence of new research on the more fundamental areas of inquiry.[57]

As with the Society for Industrial and Applied Mathematics conference, Kline's letter and criticisms drew attention to the way SMSG's
textbooks provided a venue for a larger disciplinary debate about the
nature of mathematical knowledge. By the start of the 1970s, G. Baley
Price, a former head of the Mathematical Association of America and
first chairman of the Conference Board of the Mathematical Sciences,
looked back on the previous decade and concluded that "in recent times
the mathematicians have tended to divide into two groups—the admirers
of the queen and the admirers of the handmaiden." Price's quip referred
to Carl Friedrich Gauss's characterization of mathematics as "queen of
the sciences" and E. T. Bell's counter that it was also "handmaiden" to
them. Price neatly divided the mathematical world into two camps: one
that envisioned the practice of mathematics as fundamentally divorced
from the subject's usefulness, and another that believed physical applications determined the course of mathematics.[58] While clearly oversim-
plifying, he was right to point out that the very nature of mathematical
knowledge was at stake. One SMSG supporter, mathematician Albert E.
Meder, Jr., had wondered back in 1958, at the start of the controversy,
"whether in point of fact, Professor Kline really likes mathematics." He
continued, "I do not deny that he is a mathematician, and one of con-
siderable competence. I do not deny that he is a teacher of mathematics, and from the views he expresses concerning mathematical pedagogy,
I am inclined to believe that he is a good teacher of mathematics. But I
think that he is at heart a physicist, or perhaps a 'natural philosopher,'
not a mathematician."[59]

Meder's musings centered on the relationship between pedagogical criticism and disciplinary allegiances—allowing that one might do mathematics without being a bona fide mathematician. For his part, Kline suggested that the mathematicians writing curricula had themselves come from outside the mathematics mainstream and were in fact—as he put it in 1961—"men from the more remote and less scientifically motivated domains of mathematics who have felt free to devote themselves to curriculum work."[60] Such attacks bear witness to the way SMSG's curriculum revealed the tensions at the heart of the profession.

*   *   *

The new math was not about the replacement of rote learning with conceptual "understanding." No mathematician advocated learning mathematics by memorization, and no mathematician wanted students to think that the concepts of mathematics did not have important applications. Rather, the new math presented a new idea of what mathematics was and why it should be learned. "Modern" mathematics for Begle and Stone was a discipline bound only by the requirement that mathematicians reason well. Their views nicely echoed those of congressional funders and educational supporters of the new curricular developments, who had emphasized the need for a revamped math curriculum because they believed rigorous intellectual discipline had been waning in schools. The model of mathematical discipline espoused by SMSG, however, was not even universally accepted among mathematicians, let alone among the millions of students, parents, and administrators who were exposed to the new textbooks. SMSG's books, after all, simply did not look like "traditional" math books, and those who believed that placing mathematicians in charge of reform efforts would "restore" traditional mathematics to its proper place in the curriculum would eventually be deeply disappointed. Nowhere was this gap between "traditional" and "new" math more evident than in the reform of elementary arithmetic textbooks, and it was in these textbooks that the emphasis on mathematics as a mode of thought—rather than a technique of calculation—was most clearly and forcefully expressed.

CHAPTER FOUR

The Subject in Itself

Arithmetic as Knowledge

A rthur Bestor, an influential education reformer in the years just prior to SMSG's creation, repeatedly emphasized in his book *Educational Wastelands* that discipline and freedom would be intertwined in an ideal student: "His is a disciplined mind. And because his mind is disciplined, he himself is free." Bestor warned that the consequences of failing to cultivate freedom through discipline had already been sketched in George Orwell's novel *1984*.[1] In one of the pivotal scenes, the protagonist Winston faces his tormenter, and state agent, O'Brien:

> "Do you remember," [O'Brien] went on, "writing in your diary, 'Freedom is the freedom to say that two plus two make four'?"
>
> "Yes," said Winston.
>
> O'Brien held up his left hand, its back towards Winston, with the thumb hidden and the four fingers extended.
>
> "How many fingers am I holding up, Winston?"
>
> "Four."
>
> "And if the Party says that it is not four but five—then how many?"
>
> "Four."
>
> The word ended in a gasp of pain.
>
> . . .
>
> "You are a slow learner, Winston," said O'Brien gently.
>
> "How can I help it?" he blubbered. "How can I help seeing what is in front of my eyes? Two and two are four."
>
> "Sometimes, Winston. Sometimes they are five. Sometimes they are three.

Sometimes they are all of them at once. You must try harder. It is not easy to become sane."

Eventually Winston signaled his surrender by tracing $2 + 2 = 5$ into the dust on a table.[2] Orwell's novel implied that "two plus two must be four" and "two plus two must be five" were different sorts of claims. It was not just a matter of reaching the correct answer; it was a question of what sort of *person* would arrive at the answer, and the way in which being able to reason freely meant being able to reject a decree that "two plus two must be five."

This chapter uses SMSG's elementary textbooks to analyze how students were supposed to acquire the ability to calculate. There were few, if any, complaints about a growing lack of computational ability in elementary school students at midcentury—and no teachers desired students who could not add. (As SMSG's official position paper explained, "While it is possible to argue that certain conceptions in mathematics are only fashion, it is never fashionable to get the wrong answer."[3]) The elementary schools were simply not said to be a part of the manpower crisis or the crisis in education, and SMSG's books were never posited as any sort of fix. At no point did SMSG's authors promise that their textbooks would enable students to calculate more accurately or quickly. Rather, SMSG's writers were concerned that students learn to calculate in the right way.

Elementary arithmetic is a far richer and more complex subject than competent practitioners—for whom addition and subtraction are mechanical, if not trivial—often assume. Numbers are slippery objects. As early as the nineteenth century, mathematicians ceased to agree even on the basic definition of a "number," and some of the best mathematical minds spent years trying to set out an adequate explication of the fundamental nature of numbers. It was this sense of uncertainty at the core of the field that prompted one of the world's leading mathematicians, David Hilbert, to proclaim in the early part of the twentieth century that "the mathematician was thus compelled to become a philosopher, for otherwise he ceased to be a mathematician."[4] Learning to use and understand numbers entails far more than learning a particular set of facts.

Elementary arithmetic nevertheless remained an initial, and influential, exemplar of incontrovertible knowledge for children. By the middle grades, counting and arithmetical calculations were assumed as given. The elementary math classroom was consequently a setting where students learned what reliable knowledge looks like and how it could be acquired. SMSG's writers wanted to reform arithmetic textbooks because they believed changing the way students learned to add would change the way students learned to think.

The effect of SMSG's textbooks on actual students was different from that imagined by authors or critics, however. Students were obstinate; books were soiled; fire alarms and field trips altered lesson plans; and instructors sometimes only had a shaky hold on the material they were responsible for teaching. Even something as basic as the way teachers used textbooks remained a "secret garden," poorly understood and apparently greatly variable.[5] Textbook writers often operated in the realm of the ideal, imagining how the material might shape an archetypal student. In the case of math textbooks, this image entailed a genderless, raceless, classless version of the student body, a vision not tied to particular schools or individuals. Of course, this was a fiction. If a student was supposed to be trained to reason like a mathematician, it mattered that the mathematician at this time was overwhelmingly likely to be white, middle class, and male.[6] (And, that the person teaching the subject to the student was likely to be female, etc.) It is nevertheless an important fiction to grapple with, for it contained the claim that the math classroom could cultivate a particular set of mental habits.

This chapter analyzes SMSG's promotion of "right" thinking by first surveying the landscape of Cold War fears about methods of reasoning, particularly anxiety concerning an overreliance on authority. Orwell was certainly not the only person in the period who linked the acquisition of scientific and mathematical knowledge to larger epistemological claims about "free" reasoning. After a close reading of how SMSG envisioned students would ideally learn arithmetic, the chapter turns to the psychological aspects of learning math in this way. That is, to explanations of how the new math would actually promote a new way of training students to think.

Authority and Mathematical Certainty in the Cold War

From the start, midcentury math curriculum reformers were focused on the problem of authority. As Howard Fehr, an SMSG supporter and one-time head of the National Council of Teachers of Mathematics, elaborated, "It is not sufficient to have acquired only a set of facts and procedures, called knowledge, which would simply act as blinders on his mental vision. It is essential for the student of mathematics to have acquired the faculty of being able, by his own wit, to learn more mathematics, to solve new problems, to adapt his past knowledge to new knowledge and new points of view; and, above all, it is essential to him to have been liberated from the shackles of authority."[7] Indeed, the potentially disastrous effect of an overreliance on authority was of general concern in this period. One well-known and publicized example was the work of psychologist Stanley Milgram, which seemingly demonstrated the ability of authority figures to convince "ordinary" Americans to commit reprehensible acts. In 1961, Milgram explained his fear that "the kind of character produced in American society . . . cannot be counted on to insulate its citizens from brutality and inhumane treatment at the direction of malevolent authority." Similar midcentury research, such as that of the Frankfurt school's *The Authoritarian Personality* and its F-scale, warned both that ideological commitments arose from individual personalities and that authoritarian personalities were alarmingly common even in the United States.[8]

Milgram and *The Authoritarian Personality* were part of a broader context of anxieties about authoritarian influence and the ability of Americans to remain intellectually resilient. After the Korean War ended, nearly two dozen American prisoners of war initially refused to be repatriated, provoking fears about the dwindling strength of the American character. As word of apparent brainwashing—or at least collaboration with the enemy—spread, Americans began to speculate that something must have been wrong with the soldiers' intellectual training. Army psychologist William Mayer spoke for many when he suggested "pampered" American youth were unable to withstand the stress of war. Both "dependent personalities" and "authoritarian parents" were blamed for the defections, and the curriculum was indicted in contradictory ways: schools had provided either too much radicalism or not enough. While scientists and journalists debated the causes of the apparent defections, the episode was widely read as evidence of the dangerous deficiencies of the American intellectual character.[9]

Much midcentury research underscored the relationship between types of people and reasoning abilities. The Cold War brought to the forefront anxieties about, and theories of, the relationship between forms of thought and social order. Social scientists in particular focused on claimed differences between Soviet and American personality types. The Soviet personality was said to be inflexible and rigid, characteristics that ought to be studiously avoided. Descriptions of the "good" American mind emphasized its "inner-directed" and creative nature.[10] As the Korean prisoner scandal suggested, however, the ideal American personality always involved balance—students needed an open and free mind, but still had to make the right decisions.

Many intellectuals in the 1940s and 1950s worried about the role of free or democratic inquiry in fields like mathematics and science where questions seemed to have only one right answer. Certain practices could be both constrained and creative, from the scientific "norms" of American sociologist Robert K. Merton to the conception of "normal science" popularized by philosopher-historian Thomas Kuhn.[11] The Hungarian-British scientist Michael Polanyi likewise argued that the objectivity and power of science, and its universality, arose from the free expression of individual rationality. As Polanyi described the paradox (in terms later echoed by Bestor's invocation of Orwell), "The premises of freedom will thus be secured by compulsion."[12] If scientific thought took place at the juncture of freedom and necessity, then any attempt to teach scientific reasoning would have to grapple with this fact.

In the decades leading up to midcentury, philosopher Ludwig Wittgenstein suggested that the teaching and learning of mathematics perfectly embedded this tension between creativity and coercion.[13] Based on his own experiences as an elementary teacher, Wittgenstein posed the hypothetical question of continuing a sequence, 2, 4, 6, 8, . . . or of answering the question $57 + 68 = ?$ In either case, students needed to logically and creatively apply past experience to new situations.[14] Wittgenstein used these scenarios to make a larger point about authority. He suggested that with a problem like $57 + 68 = ?$ students would not, on their own, be able to distinguish between addition and some other practice called "quaddition," both of which had been exactly the same on *every* example heretofore, but now diverged on this particular question. Harry Collins later called this the case of the "awkward student": a student who resists being able to "go on" correctly after receiving seemingly adequate rules or examples.[15] Wittgenstein hypothesized that "pointing out the mistake" might *always* boil down to a teacher saying "that's just the way it is" and thereby make mathematics into a capricious activity, based on the whim of the teacher. If mathematics were truly to exemplify the use of past knowledge to acquire new facts, how might a teacher or textbook ultimately model the process of freely arriving at certain knowledge?

SMSG's writers, while never referring directly to this question, still had to effectively answer it. And the answer certainly wasn't to encourage students to blindly defer to authority. This was the sense in which reformers like Fehr and SMSG wanted to "liberate" students from the "shackles of authority." As an SMSG-allied reform group described the goal in 1963, "whenever possible, the child should have some intrinsic criterion for deciding the correctness of answers, without requiring recourse to authority."[16] If the algebra and geometry textbooks were models of the kind of mathematics SMSG wished to convey, then the elementary arithmetic textbooks exemplified how mathematical knowledge might teach students to reliably arrive at certain knowledge generally. That is, how to know something authoritatively without recourse to authority.

SMSG and Elementary Math

The new math was not supposed to involve the elementary schools. In creating SMSG, the National Science Foundation explicitly restricted its activities to grades 7–12.[17] Partly political—forays into the high school curriculum were already considered vulgar enough for the nation's scientific elite—and partly practical—few practicing mathematicians claimed to understand the nuances of teaching seven-year-olds—the NSF curriculum content programs were not meant to reform elementary arithmetic.

Nevertheless, within months of beginning operation, Begle expanded SMSG's purview to include the creation of textbooks for every grade. A key instigator behind the expansion was Marshall Stone, who had been the most forceful advocate for basing reform on a "modern" conception of mathematics. By July 1958, during the very first writing session of SMSG, Stone and Begle exchanged letters discussing the possibility of a reform of the elementary curriculum. Begle proposed a conference the following year to investigate what, if anything, SMSG should do about the primary grades. At this subsequent conference, attendees unanimously adopted the recommendation that the elementary grades be examined and a new curriculum constructed for them. The Advisory Committee later agreed that SMSG should take on the project directly rather than delegate it to teachers' organizations or other curriculum groups.[18]

SMSG began its elementary work for a variety of reasons. First, the early 1960s marked an important institutional change for the young organization. In February 1961, noting that SMSG had already produced new textbooks for the secondary schools and monographs on the relevant mathematical topics, Begle convened the Conference on the Future Responsibility of School Mathematics. The original work SMSG had set out to do had largely been accomplished, and Begle sought advice on the organization's direction. After the conference, SMSG's leaders decided to shift the group's attention to the elementary school curriculum and to new initiatives to test the efficacy of curricular materials. The group received a grant in excess of $8 million from the NSF in June 1961 (equivalent to over $62 million in 2013), supporting these new initiatives—about double what SMSG had spent accomplishing its initial goals.[19]

Even before receiving the grant, the organization had begun work in earnest on grades 4–6, producing units for those grades in the summer of 1960. By the summer of 1962, SMSG was developing K–3 textbooks while testing the material for grades 4–6. It took another two years for the entire K–6 series to be widely available.[20] Elementary teachers were less involved than secondary teachers had been, and consequently the makeup of writing teams was skewed toward mathematicians. The official group for the first grade textbook, for example, included four professors of mathematics, two professors of education, and only two representatives from school districts (and not necessarily current teachers).[21]

Second, Begle had accepted a post at Stanford University in the fall of 1961 and took the entire curriculum project there with him. Symbolically, the move was significant because Stanford had a strong emphasis on the study of child psychology and because Begle was hired in the school of education, not the mathematics department. (In fact, Begle's shift into education was seen by mathematical colleagues as decisive: his request for a joint appointment in the math department was rebuffed, apparently to his great dismay.)[22] SMSG would continue to be run by someone with a doctorate in mathematics, but both Begle and SMSG were increasingly affiliated with professional educators, and in particular educators with an interest in the elementary grades.

Third, and most importantly, SMSG wanted to take on an elementary project for pedagogical reasons. It was precisely the arena where the group could most affect a student's ability to learn what counted as reliable knowledge. Reformers lamented that the elementary school curriculum was considered entirely separate from that of the high schools— SMSG believed it was untenable to reform algebra and geometry textbooks while maintaining rote learning of arithmetic. Instruction in arithmetic, as the introduction to mathematical concepts, should also be connected to the underlying nature of mathematical knowledge.[23]

SMSG's textbooks did not neglect the role of computation but presented it as a consequence of the structure of mathematics. As Stone explained, "our objective must be the double one of developing purely technical skills (e.g., in the art of computation) and of preparing the way towards mathematical insight into the relations which give arithmetic and elementary geometry their characteristic structures."[24] The inservice course to train elementary teachers on SMSG materials was even more explicit: the textbooks would first introduce key concepts and ideas as the "necessary foundation" for the "related development" of "appropriate skills" and the "ability to use mathematics effectively."[25] The elementary books should not just develop technique but also teach the basic "relations"—structures—underlying the entire field.

First graders would start the year off in SMSG mathematics not with the symbols of arithmetic but with representations of sets. The writers envisioned students spending the first weeks simply manipulating sets of objects, physical and abstract: the set of all students whose names begin with L or the set of animals that live in the local zoo. The point, the authors emphasized, was to teach the ability to distinguish elements that belonged to a particular set from those which did not. Then, students learned to distinguish between a set, e.g., {girls whose names begin with L}, and members of that set, e.g., Laura. One object might belong to many different sets, and some sets could be clearly defined but have no members, e.g., {months with 38 days}.

After students had become familiar with sets, they would be asked to compare and order them. Teachers were instructed to have students match up elements from different sets and to discuss whether members were "left over" after pairing. After matching, if one set had elements left over, it would be labeled "greater," and the other "less." Students also learned to recognize "equivalent" sets, meaning sets that had no extra members after matching. Comparison through matching allowed the concept of order to be introduced: one could rank sets by size. All these tasks were carefully presented without counting or numerical represen-

Are the sets equivalent?

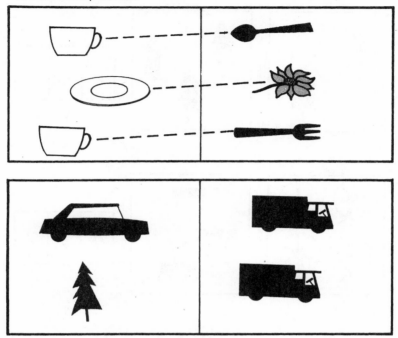

FIGURE 4.1. Students were taught how to pair elements to determine if sets were equivalent. SMSG, *Mathematics for the Elementary School*, 1.

tation. While acknowledging some students might come into the class already knowing how to count, textbooks instructed teachers to direct the students to match, not count.

The writers then made the concept of *equivalence* the crucial link between sets and numbers. Teachers told students that equivalent sets share a property, and this sameness was called a "number." Each number, in turn, was associated with a numeral—e.g., 2—which was a symbolic representation of the property of equivalence that characterized certain sets. Numbers do not belong to one set, or to particular elements, but rather to a relationship between sets. Students now faced the task of looking at a picture of objects and answering the question, How many?

An operation of arithmetic was taught as a process with sets. Addition was a name for the act of "joining" sets. Students would be shown a set of flowers, and a set of wagons, and then shown how a third set was

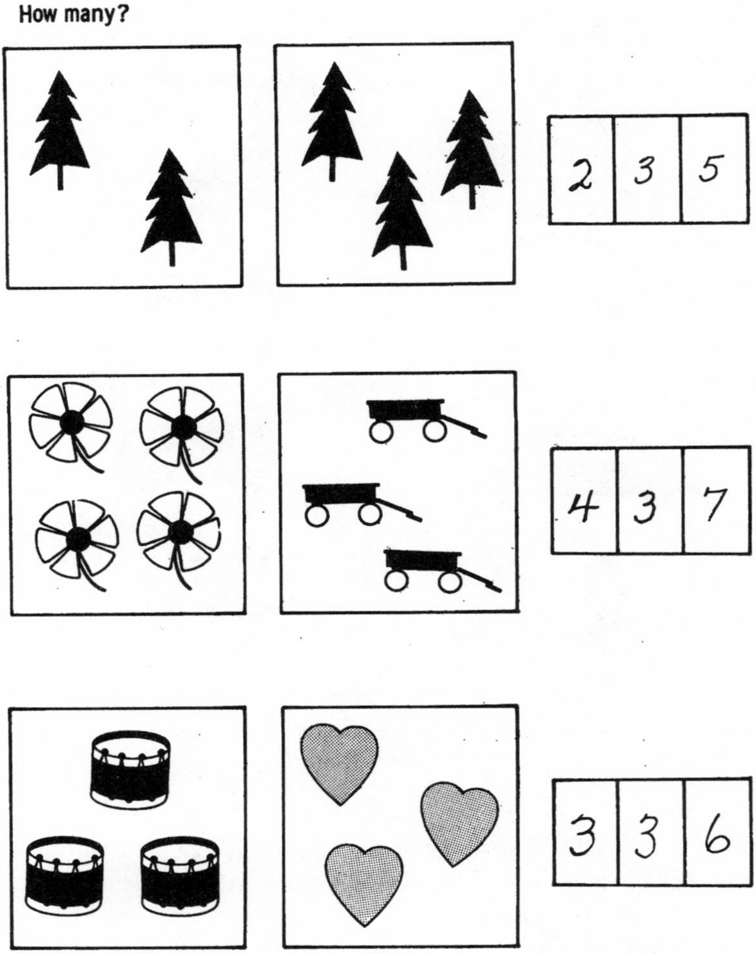

How many?

FIGURE 4.2. Labeling sets with numerals was the first step toward the more familiar version of addition. SMSG, *Mathematics for the Elementary School, Teacher's Edition*, 123.

the set of flowers and wagons. Then, teachers asked "How many?" for each of the three sets.

Sometimes the sets to be joined had the same kind of element—trees joined with trees—and other times had different kinds—wagons and flowers, for example. The process of joining did not have the usual trappings of addition at first—the plus sign or the equal sign. Rather, the emphasis was on labeling sets with numerals.

Sets were similarly used to define other operations. Subtraction required the partitioning of sets; multiplication implied the grouping of sets; division entailed the partitioning into sets of sets. Concepts of order and operation were introduced first by referring students to sets of objects and then showing how the usual symbolic representation was a means of labeling and recording operations with sets.

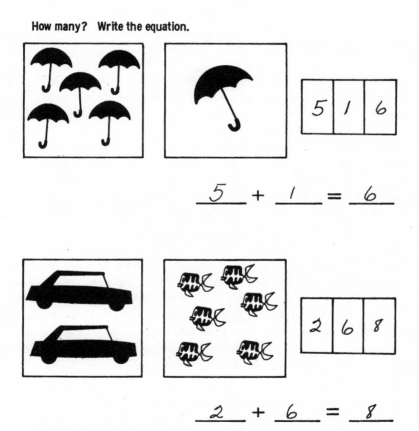

FIGURE 4.3. Only after weeks of working with sets did students arrive at the traditional symbolism of arithmetic. SMSG, *Mathematics for the Elementary School, Teacher's Edition*, 127.

In this way, students would learn to comprehend $5 + 1 = 6$ by knowing that a set has six members if it is the union of two sets that have five and one members, respectively.[26] The $+$ symbol stood for the process of joining sets, while the $=$ sign stood for the usual equivalence relationship—which is just to say that $5 + 1$ and 6 were different names for equivalent sets. Only at this point would students finally be able to do familiar arithmetic problems—the sort of problems they had perhaps expected to find in math class all along. SMSG's goal was not to increase computational speed or accuracy; it was to have students learn arithmetic in the "right" way.

SMSG and the Psychology of Arithmetic

Begle recognized that designing a new elementary curriculum concerned more than shifting the writing teams from algebra to arithmetic. As he explained to another SMSG board member, "Problems in [the] first three grades are more psychological rather than mathematical."[27] Most traditional math textbooks had been written with the assumption that elementary students did not have the mental capacities to handle anything much more complicated than memorization and basic symbolic manipulation. Could students actually learn using SMSG's texts?

The new math was obviously not the first occasion in which textbook authors confronted the psychological aspects of learning arithmetic. John Dewey, when a young philosopher-psychologist at the turn-of-the-century University of Chicago, complained that teaching arithmetic either as the memorization of ciphering rules or as the inscription of empirical investigations would be ineffective. Dewey, whose work would soon become foundational for progressive educators, wrote one of his first educational monographs on the role of psychology in mathematics education. In his book *The Psychology of Number and Its Applications to Methods of Teaching Arithmetic*, written with James A. McLellan, a Canadian professor of education, Dewey wanted teachers to present arithmetic in a way consistent with the "psychical" origin of the number concept. *The Psychology of Number* emphasized that learning arithmetic was simply an instantiation of acquiring reliable knowledge in general. Dewey's later philosophical work, including his popular *How We Think*, explained that logical thinking was "habitually exhibited" by the synthesis and analysis of number. McLellan and Dewey ultimately wanted to instill mathematical reasoning as "a normal habit of intrinsic mental working."[28]

SMSG also had the same desire to instill mathematical reasoning as a model for general reasoning, but Begle found it difficult to locate enough people with relevant educational expertise to update SMSG on the current opinions of psychologists concerning the teaching of arithmetic. As Ohio State psychologist Sidney Pressey wrote to Yale's Claude Buxton, educational psychology had few practitioners and even fewer young psychologists interested in it. Lee Cronbach of the University of Illinois concurred: "I do believe that half the psychologists in the country are spending their time writing letters back and forth nominating each other for responsibility in connection with the improvement of education. There are precious few people who are known to be interested and qualified." Eventually Begle signed on a group including Richard Alpert of Harvard, L. W. Doob of Yale, and Edward Swanson of the University of Minnesota.[29]

SMSG discovered that even if psychologists could be found, they did not seem to be able to arrive at a consensus concerning the nature of learning. Stone said as much in the initial conference on SMSG's involvement with elementary curricula. Recognizing the need to incorporate psychological understandings, Stone nevertheless concluded that it was not clear at present how to do so: "While psychologists can provide much more information about the unfolding of the child's mind than they could twenty-five or fifty years ago, it seems clear that they are only on the threshold of an understanding of the problems of mental development and concept formation." (Stone's views were largely consistent with psychologists' own assessments of the state of the field of developmental psychology.) Begle himself was skeptical of the value of psychologists, privately believing them to lack rigorous methods. Although psychologists played an important role in this initial conference and were consulted periodically, SMSG followed the lead of its mathematicians.[30]

SMSG could justify this decision in part because of Harvard psychologist Jerome Bruner. In September 1959, the NSF sponsored a conference on education in the sciences at Woods Hole.[31] Directed by Bruner, the conference brought together many of the influential curriculum designers and scientists of the period to discuss the underlying psychological and educational principles of their innovations. Representatives from SMSG—including Begle and Stone—joined with representatives from other curriculum projects. In Bruner's summary of the conference, the participants acknowledged that the notion of "formal discipline"—sometimes known as "faculty psychology"—had been largely discredited. This theory held that particular pedagogical activities could strengthen skills that might then be used in a variety of contexts (e.g., learning Latin grammar might enable students to process complicated information more efficiently in a business setting). Bruner did, however, claim that general "transfer of training" could still occur in classrooms. Appropriately designed curricula might enable "massive general transfer" in students by teaching them to "learn how to learn."[32]

The key to transference was the idea of structure, which Bruner defined vaguely as the way "things are related." "The teaching and learning of structure," Bruner said, "rather than simply the mastery of facts and techniques, was at the center of the classic problem of transfer."[33] Bruner claimed that knowledge could transfer to other domains if the focus of pedagogy was on teaching the structure of knowledge rather than its content.

Bruner's work drew heavily—if idiosyncratically—on the research of Swiss psychologist Jean Piaget. Bruner claimed that Piaget's research had shown that the mind itself was structured in a particular way. Writing later, but citing evidence known to Bruner, Piaget explained his conclusions: "there exists, as a function of the development of intelligence as a whole, a spontaneous and gradual construction of elementary logico-mathematical structures and that these 'natural' ('natural' the way that one speaks of the 'natural' numbers) structures are much closer to those being used in 'modern' mathematics than to those being used in traditional mathematics."[34]

Piaget posited discrete stages of intelligence, which corresponded to the structures of modern mathematics. It was not that the mind was mathematical but that the characteristics of certain mathematical structures and operations were homologous with the characteristics of certain mental stages. The "central problem of mathematical teaching," according to Piaget, "is that of the reciprocal adjustment between the spontaneous operational structures proper to the intelligence and the program or methods relating to the particular branches of mathematics being taught."[35] Mathematics needed to enable students to connect what they were learning with the way their minds were already structured. For Piaget, "modern" math—the structures and systems animating research-level mathematics—was much better suited to the task than traditional math—the math of memorized facts.

Bruner took Piaget's research as evidence that mathematics could be taught at a younger age because it was amenable to the underlying process of development in the brain. In particular, Bruner believed that "modern"—axiomatic and structural—topics in mathematics were most appropriate for teaching because "the sequence of psychological development follows more closely the axiomatic order of subject matter than it does the historical order of development of concepts within the field." Bruner claimed that if mathematical structures were taught in the right way, they might speed up intellectual development and facilitate the learning of other domains. Piaget himself had been skeptical about the transfer of increasingly sophisticated mathematical ideas to the elementary schools—his research suggested the structure of the mind developed in stages of relatively fixed length. In contrast, Bruner concluded that "any subject can be taught effectively in some intellectually honest form to any child at any state of development."[36]

Bruner's emphasis on the role of structure for the "transfer of training" provided SMSG psychological justification for forging ahead with curricular reform.[37] The new math's claim to developing general habits of thought was based in the teaching of mathematical structures. SMSG posited its emphasis on the structure of arithmetic not only as an accurate portrayal of mathematics but also as the natural way to teach the subject.

Bruner, in effect, encouraged SMSG's writers to proceed without worrying too much about the psychology of learning. If anything, students' abilities had been consistently underestimated, particularly in the field of mathematics. At SMSG's Conference on Elementary School Mathematics, other speakers supported Bruner's view that it was possible to teach advanced mathematics to a younger age group than previously attempted. Paul Rosenbloom reported positively on his teaching of proofs and abstractions to elementary students, while others described successful attempts to teach first graders a modified version of Euclid's *Elements*.[38] If there was little consensus among psychologists as to the underlying development of the students' mind, there seemed to be ample evidence that students would be able to handle the new concepts SMSG wanted to introduce.

Such conclusions, based on tentative ideas of what the child's mind was like, did not sit well with William Brownell, another psychologist at SMSG's conference on elementary education. Brownell had spent his career carefully researching how children actually learned mathematics. One famous study examined the use of "crutches" in calculation and whether such techniques were actually helpful to students in their learning. Brownell's research took as its subject the development of specific tools, procedures, and concepts needed to understand the learning of arithmetic. This was in the vein of what came to be called "meaningful arithmetic."[39] Begle in particular valued Brownell's work for its emphasis on careful study and experimentation.[40] Unlike Piaget, Bruner, and Begle, however, with their emphasis on underlying structures, Brownell was not interested in generating broad generalizations about intellectual development or the essential nature of mathematics and was certainly not sold on SMSG's project. Past instruction mattered far more, Brownell suggested, than did any "mental stage" of development.

Brownell's conference talk cautioned against assuming that children might be able to handle more advanced mathematics. He warned that Bruner's claims about the interrelationship of mathematical structures and mental development might just be the latest introduction of psychological ideas into curricular development, following "faculty psychology," "atomistic psychology of learning," and the "child study movement." Brownell decried SMSG's lack of actual research into how children really learned.[41]

That Brownell's warnings went unheeded confirms the centrality of mathematics itself to SMSG's version of curriculum reform. The group posited a singular, underlying structure of mathematics as the crux to understanding the field and developing "right" thinking generally. SMSG's writers believed the structural nature of mathematical knowledge was so important that one ought to at least try to teach it to children. SMSG's overall approach, as remembered by one of Begle's students, was "if you get the mathematics right, and get it organized right, then kids will learn." While suggesting that their assumptions about students' learning capabilities would ultimately need to be tested, SMSG's leaders took Bruner's assertion that elementary students could handle the rigors of abstract mathematical ideas as a postulate for their own work.[42]

SMSG could have taken a different path. Its writers could have ignored arithmetic and the psychological and practical challenges elementary schools presented. They could have simply used students' computational knowledge—however acquired—to introduce the idea of postulates, systems, and mathematical reasoning into secondary school textbooks. They could have focused on making sure students simply "understood" the mechanisms behind the traditional algorithms, thereby making them "meaningful." Instead, they focused on introducing an entirely new way of conceptualizing arithmetic, based on a claim about the underlying structure of mathematical knowledge and its relationship to reliable reasoning in general.

The relationship between the practice of mathematics and the practice of reasoning was made explicit in an SMSG *Teacher's Commentary*— that is, in one of the supplementary guides that were meant to explain the new material to teachers. Students "come to us," the commentary suggested, "with a miscellaneous hodgepodge of disjointed facts and pseudo-facts," and it is the teacher's job to "straighten out their ideas, to build a reasonable conceptual structure upon which they can hang new facts, to distinguish between that which is significant and that which is not, and, perhaps most important of all, to understand how new knowledge is acquired."[43]

Mathematics could "straighten out" students' ideas, however, only if taught correctly. Mathematics must be shown to exemplify "clear, logical reasoning" and should be used to develop in students "an ability to transfer this type of reasoning to other situations." "The acts of naming, classifying, and generalizing are conceptual in nature. . . . The real world—whatever that may mean—reaches us only by constructing a conceptual world to correspond to it." General knowledge, SMSG's writers claimed, requires a conceptual scheme, a structure, in order for it to be understood. The authors claimed that two planes of reasoning existed: a "p-plane" of "raw data from which our theory will be abstracted," which is "primary" and "intuitive"; and a "c-plane" initially "empty," "waiting to be filled with the concepts and relations which we construct." First, students must choose axioms, ideally making them consistent and economic, simple and elegant. The axioms should be categorical in the sense that every interpretation of the system should be equivalent—as was exemplified by the geometry and algebra textbooks, which showed students how "to solve geometrical problems by means of algebra, and vice versa." Although there is no reason the p-plane need correspond at all to the c-plane, "we hope eventually to set up a correspondence," a desire that "guides our constructions and choice of language." Assuming such a correspondence, experimental or observational findings in the p-plane can suggest relations to explore in the c-plane. Only formal proof in the c-plane can establish "complete certainty by logical deduction from the axioms."[44]

The process of learning math was meant to cultivate students' ability to translate the messy world "out there" into mathematics, to reason to conclusions, and then to translate those conclusions back into practical action. As one parent reiterated when "explaining" the new math's significance to other parents, "children trained in New Math should become adults who—even if they never use numbers except to balance the checkbook or add up the grocery bill—will be able to thread their way clearly through mazes of complex ideas."[45] SMSG's goal was nothing short of using mathematics to teach students to reason in the right way.

The Virtues of the Mathematical Mind

The grounding of intellectual discipline in mathematical training was present in all of SMSG's books, but two aspects of the elementary school version are notable. First, SMSG's elementary program emphasized how democratic the group's vision was. By positing that all elementary students should learn the same version of mathematics, SMSG's plans for mathematical training ultimately implied that higher-level, structural thought should not be limited to the more "capable" students.[46] Along these lines, SMSG eventually even targeted a group as far removed as one could imagine from the original mandate of college-capable secondary students: "culturally disadvantaged" elementary school children. Here, the mission to train *all* citizens to think well through mathematics was most evident.

The inservice course SMSG produced for primary school teachers included a discussion of "disadvantaged" students and the way in which mathematics might train them to think more clearly. A first chapter, titled "Description of Culturally Disadvantaged Children," suggested that such students consisted "mainly of urban slum-dwelling people" whose problem was both low economic status and a lack of participation in "middle class culture." Consistent with the near-contemporary conclusions of sociologists of education, especially those of Basil Bernstein, the text noted that one significant obstacle in disadvantaged households was the lack of complex or abstract language.[47] Mathematics could play an ameliorating role here, as a mechanism by which one learned to distinguish the "essential" attributes of objects and to exhibit abstract "classificatory behavior." Alert teachers could provide "disadvantaged" students with the skills required to distinguish similar concepts from each other and develop more precise language.[48]

The books for "disadvantaged" students had no different material, for proper mathematical habits of thought were still desirable; the books simply moved at a slower pace. In twentieth-century American society, one did not need to use rote exercises in math class to train students for routine work. G. Baley Price, a University of Kansas mathematician and supporter of SMSG, noted that the explosion of mathematics research in the twentieth century coincided with a revolution in machines and automation. The technological world of postwar United States made obsolete the skills that the manual world had required decades earlier. Training was now a matter for the head, not the hand, regardless of the student's abilities.[49] The modern world required fewer Americans willing to work on assembly lines but more people able to creatively solve complex problems. Mathematics offered a mechanism for providing all students the necessary intellectual training, starting in elementary school.

Second, SMSG's elementary work helps to distinguish the program from other contemporary NSF curriculum reforms. SMSG engaged directly with elementary students, which the physics and biology programs did not do on the whole, and SMSG did so explicitly on the basis that students needed to be able to think mathematically, regardless of their career prospects. SMSG portrayed mathematics as the key to flexible, creative thought grounded in logical rigor. Like other reform groups, SMSG certainly wanted the lay public's view of the discipline to change.[50] Learning mathematics already counted as learning to think, however; it was simply a matter of making sure it was taught such that students learned to think "in the right way." If the goal had been to train more mathematicians or reshape public understanding of math, SMSG might have limited its involvement to secondary school programs. Changing a nation's intellectual habits meant changing the textbooks in as many classrooms as possible, starting as early as possible.

A more striking difference with other National Science Foundation curriculum projects was that SMSG portrayed mathematical reasoning as the only road to certainty. SMSG's authors explicitly criticized an overreliance on inductive practices like those in the sciences. With induction, certainty was not a guarantee; SMSG's math students were told induction "should be used with caution."[51] Even if an experiment or demonstration implied a conclusion, it should be verified by mathematical reasoning to establish its trustworthiness. SMSG suggested teachers relate a story to their math classes: "Tell about the native in China who for seventy years, upon awakening in the morning, noted that the first person he saw was another Chinese. On the eve of his seventy-first birthday, after having made the same observation approximately 25,915 times, he went to bed certain in the knowledge that the first person he would see in the morning would be another Chinese. It was a Russian!"[52]

This was obviously a Cold War joke, but also an indication that reliable knowledge was understood to be at a premium at midcentury. The emphasis on structures simultaneously solved two problems for students, according to SMSG. It enabled them to learn to calculate without relying solely on authority or established tradition. And it emphasized that reliable knowledge could only be established through integrating facts into a mathematically based structure. Scientific induction was useful, but not good enough for establishing secure knowledge.

SMSG's writers insisted upon a view of mathematics not as an exemplar of a complex set of interdependent facts to be memorized (like Latin grammar) or an inductive process by which experience was measured (like chemistry), but rather as an active process of organizing, categorizing, and exploring the structure of the world. Sets were not introduced on day one because students should know the particular elements of set theory. It was not, that is, a program of rigorous definition according to the most precise and latest mathematical research. Rather, sets provided a procedure—a practice—for understanding the structure underlying both the usual arithmetic operations and rigorous thought generally.

There is undoubtedly a tension at the heart of the new math project. Many reformers were motivated by the possibility of teaching "modern" mathematics, the mathematics of structures and symbolic systems. The point of the reforms was to cultivate habits of thought more consistent with the image of "open," creative American minds than with that of "closed," inflexible Soviet minds. As the last chapter showed, however, some mathematicians criticized SMSG's portrayal of the subject as antidemocratic, suggesting that it was supported by only a minority of working mathematicians. SMSG's praise of Bourbaki and the related metaphor for mathematical development—the bulldozing of traditional arrangements in the name of modernity—also gave reformers an authoritarian tinge, especially in the implicit reference to the autocratic Haussmannian "modernization" of Paris and to Bourbaki's militaristic namesake. And changing the portrayal of mathematics from a set of facts to an axiomatic structure certainly did not guarantee it would be taught any less inflexibly or autocratically. Nevertheless, the reforms were supposed to *reduce* students' reliance on authority.

Any such irony or incoherence was not discussed by SMSG or its supporters and was perhaps overridden by faith in the presumption that "modern" mathematics instantiated and cultivated desirable mental habits. As the curriculum made its way from SMSG's writing teams to thousands of American classrooms, it would be sold simply as the intellectual training most appropriate for the challenges of the midcentury world.

CHAPTER FIVE

The Subject in the Classroom

The Selling of the New Math

The new math affected the way millions of Americans learned mathematics. This isn't a trivial point. Wholesale, national curricular changes in the United States are, after all, rare. In coordination with state agencies, local districts controlled budgets, set teacher standards and training, and maintained oversight of the schools. Teachers, principals, district curriculum supervisors, and professors of education all played roles in the development and adoption of school curricula, and indeed were critical to the spread of the new math. While it was always easy to talk about reform, such complex institutional arrangements ensured that actual school practices were exceptionally difficult to change.[1] It took a concerted effort for the new math to emerge.

This chapter analyzes the way the new math was disseminated, focusing on the bureaucratic and political machinations of midcentury schools as well as on the rhetoric deployed to sell textbooks. The first section explores the close connection between high school teachers and SMSG writers as established through writing sessions and testing centers. In the early 1960s, new secondary school books rapidly entered U.S. classrooms, books that were largely ignored by parents and welcomed by many high school math teachers. SMSG was ideally situated to play a critical role in the promotion and dissemination of new textbooks at this level.

The second section follows the entry of the new curriculum into thousands of elementary classrooms, as school boards approved the purchase of new books and abandoned the traditional emphases of math class. *The situation in the elementary schools was markedly different from that of the high schools.* First, primary school teachers were generalists who had almost no role in selecting textbooks and usually little prior education in mathematics. An onslaught of new "modern" textbooks from publishers in the 1960s—often heavily promoted by state school boards—ensured that many teachers faced the prospect of teaching from a book about which they knew little. Second, SMSG was behind the curve in the elementary schools. The group had few institutional connections with elementary teachers, and its own model textbooks often came out *after* commercial textbooks had circulated. The experience of the new math in the elementary schools was not one dictated by SMSG. Third, entry of the new math into elementary schools occurred slightly later than its entry into secondary schools, reaching its peak in the midst of liberal plans for the Great Society. By the second half of the decade, the role of scientific manpower concerns had decisively shifted in public discourse, making way for a new emphasis on social reform and domestic tranquility.

The analysis of the logistics of curriculum development and deployment in the first few sections is followed, in the final section, by an examination of the rhetoric justifying curricular reform. The new math was sold on the promise that a new form of mental discipline was required for U.S. citizens facing an assortment of political, social, technical, and moral quandaries in the 1960s. Creators of the new math, that is, did not promise higher test scores. Advocates and eventual promoters of the materials understood the design of the new math to be as much about the politics of the American mind as about mathematics. Despite the very different trajectories of the new math in the primary and secondary schools and the gradual decline of SMSG as a driving force, the claims behind the new math remained consistent. Evaluation of the math curriculum entailed value judgments about forms of intellectual discipline. The rapid dissemination of the new math speaks to the persuasiveness of the idea that math class could, and should, shape mental habits.

A Collaborative Enterprise in the Secondary Schools, 1958–65

The deployment of new math textbooks in secondary schools was a resounding success. Within a few years of SMSG's founding in 1958, thousands of secondary schools and teachers adopted new curricula for math classes and attended training institutes to learn how to teach the revised material. In 1964–65 alone, nearly 40 percent of science and math secondary-school teachers had been involved with some sort of NSF-sponsored training.[2] SMSG's own textbooks had been utilized in nearly every state, and demand for them continued to rise. The most success-ful commercial high school series was closely aligned with SMSG ideas, and drew writers and advisers from SMSG ranks. Had the reform efforts ended after SMSG finished writing high school textbooks, the new math might now be remembered as an overwhelmingly successful reform of secondary school mathematics.

Begle initially characterized SMSG's curricular work as a specific re-sponse to the demands of secondary school teachers. New materials were needed as models not only to show how textbooks should be written, but also to keep the interest of teachers engaged in "improving and mod-ernizing their courses."[3] Begle introduced his project to a national au-dience of teachers in the *Mathematics Teacher* in December 1958: "it is the considered opinion of the group that if the busy and dedicated class-room teacher of mathematics is given a mathematically and pedagogi-cally sound text, she will willingly, even eagerly, teach good mathemat-ics to her students. With no such text, even the devoted teacher is all too frequently lost."[4] Although SMSG had begun as a program run by mathematicians through the National Science Foundation, Begle hoped that by involving teachers directly in the design of the curriculum, other teachers would be "convinced that they had a very strong influence on what went into it." Teachers, however, were "now carrying the burden" of making the textbooks work and would play a critical role in the devel-opment of SMSG's secondary textbooks.[5]

Teachers themselves had been involved in curricular reform prior to SMSG's arrival. The National Council of Teachers of Mathematics had long promoted the reform of mathematics teaching through its main publication, the *Mathematics Teacher*. As early as 1955, E. P. Northrup, a mathematics professor at the University of Chicago, noted that the re-cent "explosive development of mathematical knowledge" needed to be included in the secondary mathematics curriculum. He promoted the role "modern" mathematics should play even in the "conventional cur-riculum," asserting that the introduction of new topics would help return mathematics education to its traditional place at the center of the curric-ulum. Similarly, Albert E. Meder, Jr., a mathematician closely involved with the College Board's proposal for a new college-prep curriculum, promoted reform in a 1957 issue of the *Mathematics Teacher*: because "mathematics is ever growing, its subject matter can remain manageable only through the introduction of new, unifying principles and points of view. This, in a word, is what we are asking and advocating: that secondary school mathematics be taught from the point of view of, and through the use of, the unifying principles of the present day." The National Council of Teachers of Mathematics' president Howard Fehr noted in October 1957 that while he himself knew little about research trends in mathematics, he did know that they "make much of the classical treatment of secondary mathematics obsolete." The organization needed to "help teachers to help themselves to a better mathematical fare."[6] Even before the advent of SMSG in 1958, the *Mathematics Teacher* regularly introduced pedagogical innovations to teachers on the basis that the changing nature of mathematics demanded reform.

Once SMSG had begun to produce a series of widely available textbooks, the administration of the National Council of Teachers of Mathematics recognized that teachers and administrators would need far more guidance in changing their school's program than an occasional article on the reform efforts. In the fall of 1960, an officer of the Council and adviser to SMSG, Frank Allen, applied for and received NSF funding to hold a series of conferences around the country to introduce teachers and administrators to the new curricula.[7] The regional conferences enlisted mathematicians to discuss new areas of professional research as well as educators to explain the practical challenges of introducing new curricula. While technically nonpartisan in their promotion of new programs, the conferences were largely run by SMSG affiliates and promoted the group's reforms.

While the National Council of Teachers of Mathematics helped spread the message of reform among secondary-school mathematics teachers, SMSG itself worked to get new curricular materials into the schools efficiently. The sheer size of SMSG contributed to this effort, with about four hundred people from thirty-seven states involved in writing teams between 1958 and 1966, including university professors, industry mathematicians, and school teachers. Even compared to other NSF groups, SMSG was extraordinarily large and expensive. Paul Rosenbloom, one of the group's authors as well as the director of the Minnesota National Laboratory, noted that SMSG's size was critical to its mode of distribution. Writing a textbook with twenty to forty individuals was obviously a cumbersome way to proceed, but it had the virtue of simultaneously training a large group of people to use the textbooks, which aided in the material's wide dissemination.[8]

Perhaps the most effective way SMSG promoted its new materials was through its extensive network of testing centers. They ranged from formal research settings like Rosenbloom's laboratory to school districts willing to serve as experimental sites. These centers were not merely sites for tweaking texts and manuals, although they served that purpose. Testing centers were a key distribution method for the novel curriculum.

In some cases, like that of the public schools in Newton, Massachusetts, testing centers were located in districts with extensive experience of curricular reform. The principal of Newton High School, Harold Howe II, later became U.S. commissioner of education and an important figure at the Ford Foundation and Harvard Graduate School of Education. In October 1957, *Time* magazine listed Newton High School as one of the nation's thirty-eight best high schools based on its National Merit finalists. Two years earlier, the high school had embarked on a nationally recognized plan to increase educational opportunities for students. Through the 1950s, few would conclude that Newton's science and mathematics programs needed revision—its students, for example, placed first and second in the Westinghouse Science Talent Search in 1957. The school had long been known for scholastic excellence and commitment to cutting-edge reform.[9]

Wealthy, progressive districts like Newton became important initial sites of dissemination for the novel mathematics curricula. By early 1958, Newton's school committee began a series of responses to ensure that its students had the best math, science, and foreign language instruction available. Committee members introduced Russian as a foreign language in the fall of 1958, increased the number of laboratory courses and made them a requirement, and developed an entirely new junior high school science course. In November 1958, W. Eugene Ferguson, the head of mathematics at Newton, received an invitation from the head of the curriculum development program at the University of Illinois, Max Beberman, to join the reform effort. Ferguson would work closely with SMSG in subsequent years as well.[10] Starting in 1958, Newton High School served as a testing site as Ferguson and his fellow teachers adopted SMSG and a variety of other programs into their curriculum. In Newton, SMSG had a natural foothold simply by virtue of the district's long-standing commitment to curricular innovation.[11]

Most schools, however, had neither experience with innovative curricula nor Newton's wealthy and progressive base. SMSG facilitated changes in these other districts by leveraging an impressive roster of writers, including many administrators and teachers, who often later served as advocates for the curriculum. Begle's files testify to the numerous districts and schools requesting to be sites where preliminary SMSG materials were tried.[12] The combination of large writing groups and a substantial network of testing centers ensured a ready supply of people trained in SMSG's philosophy.

When a group of teachers and professors tried to convince the Denver public schools to modernize their program, for example, the reformers relied upon this network. Burton Jones, a professor of mathematics at the University of Colorado, attended the very first writing session of SMSG in 1958. The following May, Jones joined with the principal of Hill Junior High School in Denver to pitch the experimental adoption of SMSG materials in the local secondary schools.[13] They spoke to the board of education, the superintendent and staff, and the principals and curriculum coordinators of the Denver junior and senior high schools. They had already gained the support of teachers in the district, some of whom spoke on behalf of adopting the SMSG experimental books. Consistent with SMSG's general operation, a member of the teaching staff was also planning to gain firsthand knowledge that summer as an SMSG writer.

In Denver, the proposal to reform the curriculum emphasized far more caution than would have been required in Newton; this was not to be a crash-course effort. Jones assured the assembled parents and teachers that there was no single best curriculum but it was important to test out the new books and courses. The reformers tried to assuage doubts by citing the availability of texts, the location of a nearby SMSG experimental and in-service training center in Boulder (presumably run by Jones himself), and the extra compensation for teachers to prepare for the use of SMSG materials. One teacher told the board that while he approved the testing of the new curriculum, "We in Denver are not about to abandon our present curriculum, which has produced high quality results, in favor of a new curriculum in an experimental stage."[14] For secondary schools unfamiliar with previous reform efforts, the extensive support network allowed reformers to advocate both caution and change.

In most cases, the transition to the new curricula was tentative. Eugene, Oregon, for instance, widely adopted the SMSG materials, but did so by using six different mathematics programs, old and new, in its schools.[15] Such hybrids were the norm, as districts attempted to incorporate the new programs on a small scale—usually with their better students and teachers—and gradually move away from "traditional" textbooks in favor of the newer algebra and geometry texts.

In less-populous regions, entire states could force schools to reform. Oklahoma was one of the first to deploy SMSG materials statewide. By 1959, seven of SMSG's forty-nine testing centers were located in Oklahoma, and members of the state mathematics committee were involved with SMSG's first writing sessions. The initial testing was deemed successful, and for the fall of 1960, the state mathematics committee ordered 4,000 more SMSG *Geometry* books to be tried in Oklahoma schools. Oklahoma secondary teachers took advantage of NSF summer institutes and in-service guidance for training, and by the 1961–62 school year, over 47,000 SMSG books were in use in Oklahoma schools. The following year, nearly 250,000 students used SMSG materials in the state's schools, and in the fall of 1963, the state textbook commission adopted the new books in addition to the traditional books in order to start moving every Oklahoma math classroom toward SMSG's curriculum.[16]

The Newton, Denver, Eugene, and Oklahoma experiences suggest the existence of a flexible but robust pattern. Such early-adopter schools often had multiple individuals with direct SMSG experience, tested the material initially on a limited basis and in conjunction with other programs, and relied upon the extensive resources of SMSG, the National Science Foundation, and local mathematicians to ensure teachers and administrators were adequately prepared. SMSG was more than a source of books; the organization provided initial training as well as the infrastructure for testing and evaluation. Just a year after SMSG's founding, its books had been tried by 42,000 pupils in 45 states; the next year 100,000 books were in schools, and within five years SMSG estimated that over two million students had used the new textbooks.[17]

While impressive for an initial step, this meant that the books were still used regularly by only a small percentage of the nation's students. In order to have a substantial impact, SMSG would have to change the writing of commercial textbooks. Unlike other NSF groups, which produced finished, professional-looking textbooks, SMSG chose to market ones that were intentionally shabby. They featured no color illustrations, crudely typeset pages, and flimsy paper covers. As one SMSG adviser lamented in an executive committee meeting, "we made sure our books were badly printed, badly illustrated and badly bound."[18] The focus was on creating content that could be used to train teachers and providing a model for commercial writers. The physical condition of the textbooks emphasized both that they were part of a new and serious project and that they were not in competition with the commercial publishers.

SMSG's most important dissemination effort, in addition to the expansive testing centers, was to operate the summer sessions as training grounds for future commercial textbook writers. Begle recruited teachers like Henry Swain who had commercial experience—Swain chaired SMSG's algebra writing team, having previously coauthored D. C. Heath and Company's *First Year Algebra*—but more importantly, publishers looked to SMSG conference participants as lead authors for new commercial texts. As Begle's widow recalled years later, "To be on SMSG was an invitation to get an offer from commercial publishers to write texts. And that was fine with Ed."[19] In one of the most direct translations of SMSG material, Edwin E. Moise and Floyd Downs published their *Geometry* in 1964 with Addison-Wesley, a book that remained in print into the 1990s. Both Moise and Downs were on the geometry writing team, when Moise was at the University of Michigan and Downs at East High School in Denver, and their *Geometry* was based extensively upon the writing of the original SMSG geometry textbook five years earlier. In fact, the commercial book so closely mirrored the SMSG original that Begle had to examine it prior to publication to verify that it was different enough from SMSG's version to justify standing on its own as a commercial text. Begle approved its publication but noted that the structure and design were exactly the same as the SMSG original.[20]

The most important connection between SMSG and commercial textbook writers was a Hunter College mathematics professor named Mary P. Dolciani. While her time at SMSG was relatively short—she was involved only with the 1958 and 1959 writing teams—she would be the most important translator of SMSG's program into commercial viability. Just two years after her work for SMSG, Houghton Mifflin contacted Dolciani about the possibility of being the "lead writer" for the new series they were developing in response to the demand for reforms in secondary school math, called Modern Mathematics. She accepted and went on to be involved with dozens of textbooks as Houghton Mifflin expanded its "modern" mathematics offerings to include every grade.[21]

For her first effort, Dolciani joined with William Wooton, Albert E. Meder, Jr., and Julius Freilich to publish *Modern Algebra: Structure and Method* in 1961–62. Freilich was listed only because the group was effectively replacing his existing algebra textbook with one that included "modern" ideas, but Dolciani, Wooton, and Meder were the key contributors to *Modern Algebra*. All three had close connections to SMSG: Dolciani and Wooton as former writers (Wooton would later write a history of SMSG), and Meder through his involvement with the College Board's curriculum recommendations, which served as a blueprint for SMSG work. Texts in the series prominently displayed the authors' links to reform groups, especially SMSG.

Modern Algebra: Structure and Method was the first title in what would be a long association between Houghton Mifflin and Dolciani in their Modern Mathematics series. Over decades (and even past her death), Dolciani was listed as an author of dozens of best-selling books, including *Modern Introductory Analysis, Modern Trigonometry, Modern Geometry*, and *Modern School Mathematics*. While sales figures are notoriously difficult to come by—companies keep details confidential even fifty years after publication—Houghton Mifflin's series was widely acknowledged to have captured the secondary-school mathematics textbook market in the 1960s, and the company led in sales well into the 1970s.[22]

That a commercial textbook would both sell well and receive SMSG's praise proved to be the exception. Begle had hoped that SMSG's models would quickly be adapted by commercial publishers—and promised that the SMSG books would only be available until suitable commercial replacements could be found. By 1965, Begle attempted to promote the SMSG model by writing to Austin MacCaffrey, the executive director of the American Textbook Publishers Institute (the trade's professional organization), to clarify the terms of the SMSG copyright: no one had exclusive rights to the material of the SMSG books, and no royalties would have to be paid. Begle's plea that commercial publishers could and should draw directly from SMSG's texts worked on one level. For-profit companies effectively took over the market as SMSG had hoped, and sales of SMSG textbooks dropped off sharply after 1965. Nevertheless, Begle and SMSG were never really satisfied by the quality of commercial texts. Dolciani's *Modern Algebra* and Moise and Downs's *Geometry* were, in fact, the only two books ever considered to be suitable replacements for the SMSG texts.[23]

Even with Begle's misgivings, SMSG had a profound effect on the nation's secondary school textbooks. The organizational strategy of collaboration, and careful introduction of new textbooks, worked to convince teachers and administrators to buy and implement a new mathematics curriculum. The fact that the high school math teachers' professional organization worked closely with SMSG was also important. Parents played little role, perhaps not remembering enough algebra and geometry themselves to have opinions but more likely deferring quietly to the judgment of curriculum writers, teachers, and school boards. Despite changes in millions of textbooks, SMSG noted very few complaints with their secondary school efforts.

A "Bomb" on the Elementary Schools, 1965–70

The elementary schools were another story entirely. There was no formal organization for elementary mathematics teachers, although some journals did address their concerns. As a result, district curriculum supervisors, state and local school boards, and superintendents played far more important roles in determining the elementary school curriculum and its textbooks than they did in the case of high schools, where teachers were closely involved. Relatively few elementary teachers or administrators participated in curriculum projects like SMSG, and many teachers did not know the mathematical or pedagogical justifications for reform. One of the leaders of math reform in this period noted that while about half of high school teachers had some formal training in newer mathematics materials by 1965, such training had been given to only about 5 percent of elementary teachers.[24]

These concerns were well known by those involved with SMSG. In November 1960, a little over a year after the official start of SMSG work on elementary school material, Rosenbloom, the director of Minnesota's curriculum laboratory, wrote an article, "What Is Coming in Elementary Mathematics," published in *Educational Leadership*. "We have about 900,000 elementary teachers," the SMSG-allied Rosenbloom warned the administrators who formed the bulk of the journal's readership. "With existing manpower, the task of in-service education of elementary teachers is almost insuperable." Reformers knew that without in-service programs to adequately prepare teachers, results with the new textbooks would be poor.[25]

The very nature of elementary school teaching presented an obstacle. Harold Bacon, a professor of mathematics who served as an instructor for the NSF training institutes, warned that the average tenure of a contemporary elementary teacher was only about five years. The "average" elementary teacher, in fact, was a woman under forty who had been teaching for fewer than ten years and who belonged to no professional teachers organization. She would have observed someone else teaching mathematics only a couple of times outside of her training. She had not specialized in mathematics and had spent only a minority of her instructional time on mathematics. The teacher was likely to have completed only two college-level math courses and one math education course in addition to high school algebra and geometry. It was common for teacher training programs at midcentury to require more preparation in arts and crafts, music, and health/physical education than in arithmetic. Newly certified teachers in the 1950s, at least, had to take a minimal amount of math (with little consistency, however—requirements ranged from six to forty credit hours), but most teachers were not certified, and most professors in teacher training institutes doubted that their students had enough of an understanding of arithmetic to teach the subject well in their own classrooms.[26] A new textbook that required additional mathematical knowledge would have been a substantial burden.

SMSG writers also worried that the practical challenges of reforming the elementary curriculum extended beyond the issue of teacher training. Elementary mathematics was a muddled collection of topics compared to the traditional tracks of geometry and algebra in later grades. While high schools (at least those large enough) tended to group students of similar ability, elementary classrooms featured a mix of skill levels. It was unclear how to devise a first grade book when some of the children already had a year of study in kindergarten and others had not. Textbook writers also had to negotiate the difficulties presented by the uneven coverage and depth: what might be third grade mathematics in one classroom was fourth grade material in another—even within the same district.[27]

During the early 1960s, SMSG was extensively testing its first grade textbook and discovering that even experienced teachers under close supervision had difficulty teaching the material. Teachers reported that concepts might be understood but students found words like "equivalence" to be unpronounceable. When class size drifted north of twenty-five or thirty, it was simply not possible to carefully manage the ways students actually operated with sets. Students who could already count wondered why they were comparing sets with matching instead of just saying "how big" the set really was. And, as one teacher concluded about her peers, no matter what you put in a textbook, "it is always left up to teacher judgment and many of them don't have any that is worthwhile."[28] SMSG was well aware of the formidable obstacles facing the introduction of entirely new ways of teaching arithmetic in the elementary schools.

While SMSG's writers worried about the proper mechanisms for elementary school reform, many of the nation's elementary schools were already changing their curriculum by 1965. In January 1965, the *New York Times* ran a full-page feature, complete with stories on the philosophy behind the new curricula, the experience in city schools, and a front-page article on the various supporters and critics of the changes. The paper even had "info boxes" describing Venn diagrams, axiomatic systems, set-theoretic proofs, and modular arithmetic as examples of the new "modern" content. In April 1965, *Business Week* published a series of articles on the profitability of the new mathematics textbooks for publishers. *Good Housekeeping* published summaries of the new topics, while many districts offered night and weekend courses to prepare parents to help with homework. Audio recordings were produced for evening instruction. *Parents Magazine*, which only rarely featured explicit discussion of curricula, had a story in the fall of 1965 on parents' use of new math ideas to prepare children for what they would face in elementary school. *Harper's Magazine* reviewed some of the many books marketed as primers for the new math topics, from *Understanding the New Math* to *A Parent's Guide to the New Math*.[29] Between April 1964 and November 1965, Charles Schulz made the new math a topic of conversation in the comic strip *Peanuts*—and emphasized the centrality of intellectual habits for the curriculum's promoters—as the kids struggled to do new math with an "'old math' mind." By 1965, the story in elementary mathematics was the new math.

It was essentially not, however, SMSG's new math. By 1965, SMSG had sold over 2.6 million textbooks, but the vast majority of these were at the seventh grade level or above.[30] SMSG had little infrastructure for

FIGURE 5.1. *Peanuts*, April 22, 1964. PEANUTS © 1964 Peanuts Worldwide LLC. Dist. by UNIVERSAL UCLICK. Reprinted with permission. All rights reserved.

testing elementary texts and did not even produce a revised elementary textbook until after 1965. Unlike the case for high school textbooks, the authors of the major elementary commercial textbook series were rarely connected with SMSG writing teams, which obviously affected the group's ability to influence the elementary curriculum.[31] Other curricular programs proved more influential than SMSG. One of the earliest successful elementary programs was the Greater Cleveland Mathematics Project, whose sales in 1962 exceeded expectations by a factor of ten. Other schools adopted new elementary materials from groups like the University of Illinois Arithmetic Project and Syracuse University's Madison Project. These efforts reflected a great range of pedagogical and mathematical approaches but were unified in their claims of novelty. Following their lead, commercial textbook publishers quickly produced new series with titles such as Modern Arithmetic and Discovery Mathematics.[32] With commercial alternatives readily available, few school districts chose to implement SMSG's books when they appeared, and sales were never as robust as with the secondary textbooks.

In any case, SMSG's elementary reforms were not introduced as a collaborative process between national curriculum groups and individual teachers and schools. Begle had been careful to avoid any appearance of impropriety with federal funds and had insisted that local districts make their own decisions about what math books to purchase. Leaving decisions to districts had worked successfully with the secondary schools, especially because of SMSG's close collaboration with teacher organizations and summer institutes. Such an approach in the elementary context, however, ensured that commercial publishers would lead the process. Even more important than SMSG's choice to stay out of district-level curriculum decisions was the role of size and scale in the elementary textbook market. An algebra textbook would be used by a portion of students for at most a year; elementary books were typically done as a series and adopted en masse, so capturing the market meant capturing five or six years of schooling for every student. As a result, state boards of education, textbook adoption committees, and textbook publishers largely controlled the marketing and sale of books to elementary schools.

SMSG's main influence on the elementary schools occurred through its work with state boards of education. SMSG affiliates advised states on curriculum guidelines for all grades even before the organization had revised its model elementary textbooks. This was part of a larger shift to effect reform through top-down bureaucratic action—a transition praised by influential educators such as Harvard's president James B. Conant. In California, for example, the state department of education gathered secondary teachers in 1959–60 who were already working with experimental material, as well as curriculum directors such as Begle, Beberman, Rosenbloom, Fehr, and others, to discuss the future of the mathematics curriculum. In the summer of 1960, the California curriculum commission, which reviewed textbooks for use in the state, appointed a special advisory committee on mathematics to meet during 1960–61 and report on how to introduce textbooks that would facilitate a shift to "modern" mathematics. Between 1962 and 1964, the state formalized its approach as one based on "strands," the ideas that should run through all the levels of mathematics education, including the elementary schools.[33]

The push for new textbooks by state boards would have been in vain, however, had the textbook industry not been producing brand new material for the elementary schools. In many regards this was the most surprising aspect of the new math's dissemination because the elementary and high school textbook industry was simply not set up for rapid changes in style or content. The industry's conservatism was a consequence of the cost of developing new textbooks and the perpetual risk that older textbooks would be favored over untested new ones. Many districts, after all, simply did not order new books when finances were tight. It was far more expensive to produce a textbook than reprint an old one, and each new title (by one estimate) had to sell at least 50,000 copies to be competitive. Despite sales of over 150 million books for a typical year in the mid-1960s, the industry had never been exceptionally profitable. Furthermore, companies often focused on elementary math books, high school history texts, or similar niches, further segmenting the competition. No textbook publisher controlled over 10 percent of the market.[34]

In such an industry, timing was everything. One contemporary industry official warned that the "mortal sin" of publishing was to fail to respond to the desire of schools: "the most damaging publishing mistakes are made by investing heavily in good ideas—not bad ideas, mind you, but good ones—that are too far ahead of their time." One company, Scott, Foresman, had introduced a new mathematics series in the 1940s and 1950s which aimed to move away from the drudgery of computation. At the time, it was condemned as too advanced and sold poorly— schools simply did not want to update their mathematics curriculum. After Houghton Mifflin produced the best-selling Modern Mathematics series in the mid-1960s, the Scott, Foresman books were condemned as "not modern enough."[35] Maintaining dominance required acute sensitivity to the fickle demands of schools.

In part, changes in the organizational structure of the textbook industry enabled its rapid expansion. While postwar textbook industry conferences had been primarily focused on specific aspects of the publishing industry, by the start of the 1960s, the American Textbook Publishers Institute began to host conferences grappling with "The Challenge of Change." One contemporary analyst who tracked the industry, Frank Redding, noted a "revolution" in the structure of textbook publishing. Redding attributed the change to two financial factors: initial public offerings of stock in textbook companies combined with a slew of mergers to drive up stock prices and make far more money available for new investments. He identified at least fourteen public offerings between 1958 and 1962 and over twenty-eight mergers in that same period. The stock offerings and mergers made capital more widely available and enabled the textbook publishers to invest in new projects and novel textbooks at an unprecedented rate.[36]

The safest way for textbook companies to make money was to capitalize on the mechanisms of statewide adoption lists. Some states were said to be "open territory," allowing any company's "bookmen" to pitch their wares to individual districts. Many southern and western states instead relied on statewide adoption committees to regulate the list from which individual districts or teachers could choose textbooks. Careful state regulation of textbooks had begun in the late nineteenth century and afforded an opportunity to control the content and style of textbooks.[37] While ostensibly intended to maintain quality across the state, the adoption lists led to fierce competition among textbook companies.

Consequently, the politics of textbook adoption ensured that a small number of states had a disproportionate influence on the content of textbooks in the 1960s. More than half of total textbook sales were in only ten states, and when those states, such as Texas and California, also had statewide adoption lists, their influence was even greater. States like California only adopted a handful of mathematics textbooks, providing substantial financial incentive to cater textbooks to its population. When a Harper and Row salesman, for example, pitched his company's new *Today's Basic Science* book to Florida officials, he estimated that nearly 200,000 sales were at stake. Indeed, after this particular successful sales pitch, a number of other states, as well as urban districts in northern and midwestern states, adopted the Harper and Row volume, transforming it into a best seller.[38] Since sales figures were largely confidential, public adoption lists became both the mechanism for entering the lucrative markets in the big states and a crucial guide for other states and companies as to the most popular titles.

In Texas, the educational bureaucracy combined with publishers' financial interests to force a shift in elementary mathematics education. In 1962, the Texas Education Agency developed a self-instruction course to introduce elementary teachers to the ideas of "modern" mathematics, especially the structure of arithmetic from an algebraic viewpoint. Although just 900 of Texas's 40,000 elementary school teachers completed the course by the spring of 1963, that November the state board of education adopted new mathematics textbooks to be used in the classrooms.[39] The state planned for teachers to introduce the new material and books for one hour a week in 1964 and then implement the full curriculum the next academic year.

By the time the vast majority of Texas elementary teachers learned about the new curriculum in 1964 and 1965, the change to "modern" textbooks had largely been accomplished. As the teachers association journal *Texas Outlook* proclaimed, "Rare is the teacher today who is not acquainted with 'new math.'" Unlike the secondary school teachers in Denver, who had the luxury of careful testing, paid in-service and summer training, and consultant mathematicians on call, the Texas elementary teachers had only a brief self-instruction course that many failed to complete. While some Texas teachers were optimistic about the new textbooks, others noted the uncertainty surrounding the rapid change. One teacher labeled the process a "leap of faith." Looking back five years later, a professor of elementary education at the University of Texas noted that elementary teachers in Texas had a "bomb" dropped on them during the 1964–65 school year: "Ready or not, the New Math was here and new textbooks had to be adopted."[40] In both California and Texas, the statewide adoption committees were pushing modern books by 1964, and elementary teachers had to use what their districts adopted.[41]

The experiences of California and Texas with the new math were consequential.[42] They had the two most influential statewide adoption lists and had moved to include "modern" mathematics in the elementary schools in the early 1960s; textbook companies raced to produce textbooks pitched to their new standards. Furthermore, once textbooks had been produced for California and Texas, publishers attempted to sell the same books in "open territory" states. Unlike in the high schools, where strong professional organizations and teacher preference drove textbook selection, in the elementary schools textbook salesmen had far more of an impact. In one survey, elementary principals in the state of Washington said the best sources of information on curriculum innovations were textbook salesmen—preferred even to district curriculum advisers. States like Alabama, for example, had not been involved in the initial push toward the new math but adopted new math books once publishers started promoting them more widely. In 1965 the *Alabama School Journal* published a special issue on the new curriculum in order to train teachers who otherwise knew little about the origins or purpose of the reforms. In North Carolina, careful discussion with teachers and curriculum designers about the development of the high school curriculum was followed by sudden adoption of new textbooks for the elementary schools.[43] With testing centers in some states serving as de facto dissemination mechanisms for new textbooks and Texas and California providing financial incentives for developing the new books, other states were squeezed into adopting the new math.

Modern Math for the Modern Student

The new math quickly became standard fare for commercial mathematics textbooks. There were nearly twice as many primary and secondary mathematics titles for sale in 1965 as had been on the market in 1958. This massive increase in available titles was primarily the result of an increase in new listings: whereas only one- to two-dozen new mathematics books appeared yearly in the late 1950s, by the early 1960s the flow increased to about one hundred new titles a year until it leveled off in the middle of the decade.[44] One observer lamented that no publisher would be so "foolhardy" as to not discuss sets extensively, for "there are too many people who judge a book by its coverage of sets." Indeed, some complained that in the rush, substance suffered: physicist Richard Feynman, serving as a textbook reviewer for the state of California in 1964, claimed that there were only a few books worthy of approval—most had simply included new mathematical terms and precisely defined concepts without consideration of the goals or needs of the schools. By 1965, the American Book Company's president summed up the rapid changes by noting in *Business Week* that "anyone not in new math can count himself out of the mathematics textbook business."[45]

The marketing of the new math in both elementary and high schools was largely not about mathematical competence per se. There was little, if anything, ever said to be wrong with the textbooks in use; they did not contain pernicious errors, and students seemed to learn to calculate competently with them. While there was always a huge range of what counted as a "modern" math book, the very notion of selling a "modern" math book was popularized in exactly this period. In 1958, there were hardly any mathematics texts labeled "modern" on the market, and only a scattering appeared over the next couple of years. Starting around 1961, however, as states like California and Texas began to adopt new guidelines and SMSG began to produce revised textbooks for high school courses, new arrivals began to feature keywords pointing to their novelty and incorporation of "modern" mathematics.[46] By 1965, nearly every mathematics title claimed to be "modern," "structural," "new," or "contemporary."

SMSG had been the group most vocally justifying the new textbooks by referencing the rapidly changing nature of the contemporary world. Almost the same language was repeated constantly through articles written by SMSG affiliates, warning that "children now in elementary school will face problems we cannot predict."[47] (The consistency of message was no accident: Begle carefully monitored the language used by SMSG's supporters.)[48] In the "Basic Principles" of SMSG's initial writing session, the very first claim was that "the world of today demands more mathematical knowledge on the part of more people than the world of yesterday, and the world of tomorrow will demand even more."[49]

In a later speech to teachers Begle elaborated on these "demands": "a profound change is taking place in our society," as "tremendous use is being made of science, technology, and mathematics, and the rate of increase of this use is itself increasing rapidly." He pointed to Margaret Mead's recent characterization that a

hundred years ago, say, it was reasonable to predict the world would not change very much in the next 20 or 30 years and that a particular skill taught in school at that time would remain useful to the student after he had left school. . . . Even 50 years ago this was still pretty much true. But let us look at the situation 20 years ago and compare it with today. Consider the mathematical skills which were taught 20 years ago. Today we can look around us and see that most of these skills, not all, but most, are still useful, but we also see that today we need many new skills, some of which were not taught

20 years ago because they were not useful then, and others which were not
even known 20 years ago.[50]

The changes in the world were wrought in part by mathematics, and
even more mathematics would be required in the future. Particular skills
could not be relied upon, for they might not be needed by the time students left school years later. The world was changing too rapidly to depend on traditional knowledge. As Begle had explained at the outset of
SMSG, learning math for "intelligent citizenship" meant learning math
in a world where no one can predict the important and useful skills of
the future.[51]

When promoting the new curriculum for teachers and administrators,
SMSG affiliates provided similar justifications for reform. One SMSG
supporter explained the justification for the new math to teachers by
claiming that the math teacher must ensure that "the members of the
next generation are to have the knowledge necessary to operate the complex civilization they inherit."[52] A professor in Missouri wrote to teachers in his state in 1962 that this was the perfect time to make the change
to a "modern" mathematics program. "You cannot walk the middle of
the road holding hands with tradition on one side and modernism on the
other. . . . You have to make a choice."[53] A professor writing in the journal for the New York City board of education used language rarely deployed in the context of mathematics education: "Every age must break
free from the swaddling clothes in which the past has bound it." Even
mathematics, the subject that still taught thousand-year-old theorems,
needed to modernize. He concluded, "Change is the norm."[54]

The growing federal education bureaucracy produced pamphlets
pointing parents to this broader role for mathematics class in a modern,
complex world. *Modern Mathematics and Your Child*, produced in 1965
by the Department of Health, Education, and Welfare, noted that mathematics meant more than "arithmetic, algebra, and geometry; it represents a way of thinking and reasoning" for evaluating knowledge generally. While the "world of mathematics" was much larger than its direct
applications, math was indeed useful for solving the "complex problems
of industry, Government, and defense." Alongside boilerplate recommendations to set aside a quiet study space and stay informed of a child's
education, the pamphlet reassured parents that the strange textbooks
should not be cause for concern, only more evidence that the "times
have changed."[55]

These claims were at once well worn and yet specific to the Cold War. Exaggerated reports of a formerly static world experiencing rapid change driven by mathematical and technological developments flourished well before midcentury—providing, for example, the basis for Charles Dickens's satire of Thomas Gradgrind in his 1854 novel *Hard Times*. The Cold War context meant, however, that the development of disciplined and transferable intellectual habits was a matter of existential importance. The National Council of Teachers of Mathematics 1959 Secondary School Curriculum Report, for instance, included an assessment from a subcommittee chaired by J. D. Williams of the RAND Corporation. The piece, "The Place of Mathematics in a Changing Society," began by reminding readers that just a hundred lifespans would take us back to the start of recorded history. Even this fact did "not reflect adequately the explosive character of the development of civilization and knowledge. . . . The rate of development during our lives has become fantastic, though we tend to become blind to it; e.g., it took us a couple of months to become blasé about man-made satellites." Unfortunately, the committee wrote, this tendency had recently become dangerous: "if some new group surges into the lead, it will not be astonishing if it happens to obliterate us in the process." In dealing with such a threat, any attempt to "extrapolate" the future distribution of occupations inevitably "brings to light a picture of alarming implications."[56] The rapid changes of the recent past meant that it was more important than ever to provide the necessary intellectual training to students.

Teachers did not miss out on the message. Donald F. Define, a high school teacher, wrote in 1962 that "the age of technology is upon us and we must face the changes which accompany it. The young people of today must be prepared to take a role in tomorrow's world. The demands being made upon the youth of today are in many ways different from those placed upon their parents."[57] Another teacher, B. R. Reardon, quoted from a pamphlet titled "The Resourceful Teacher" to inform readers of the *Alabama School Journal* of the urgency of the situation: "Men are orbiting the earth, satellites are rocketing into space, astronauts are preparing to go to the moon—all as a result of the great explosion of knowledge that is taking place in our modern era. Today's culture is a 'mathematized' culture. New and startling technological and scientific developments are occurring daily. . . . Resistance to change in such a fast and changing world is now almost ridiculous." Reardon concluded by noting that the "modern teacher, to stay modern must adopt modern

methods and take a fresh, new outlook on methods and approaches. . . .
We must be brave and bold and continue with a pioneering spirit." Reardon claimed that as the nation must keep up with technological ad-
vances, so a teacher must "fall in line with the times."[58] Math reform was
necessitated by the changing demands of the modern world.

While Reardon never actually defined a "modern teacher" or "modern methods," the use of the rhetoric of modernity had other resonances
in this period. From foreign policy to economics, in both domestic and
international contexts, the need for "modernization" was widely promoted, even if the measure of "modernity" often involved little more
than a comparison with the United States.[59] Daniel Lerner's 1958 book
The Passing of Traditional Society, for example, established one key te-
net of modernization as movement toward midcentury American cultural ideals. His survey of six Middle Eastern nations documented their
"progress" on the path to modernization, with progress measured by
levels of literacy, urbanism, media participation, and, especially, empathy.[60] Lerner's notion of empathy referred to the specific ability to imagine oneself in new and strange situations, paralleling the skills teachers
and professors needed to embrace the new math. Empathetic individuals, like modern educators, would actively welcome change rather than
passively accept traditional structures of social order.

Whether emerging from accounts like Lerner's or from theories of society developed contemporaneously by Talcott Parsons and Edward Shils,
the language of "modernity" signified claims about the nature of society.
In the case of the "modern" mathematics peddled by textbook publishers, curriculum designers, and activists, the reference was to claims about
the modern world's complexity and technological sophistication, which
presented unique challenges for the training of young citizens. Rosenbloom, an SMSG ally and director of the Minnesota National Laboratory, gave perhaps the clearest account of the relationship between curricular change and the modern world in his forward to the proceedings of
a 1961 conference, *Modern Viewpoints in the Curriculum: National Conference on Curriculum Experimentation*. He began by reiterating Begle's
belief that "our fundamental problem is that we are the first generation
in history which must educate children for an unforeseeable, changing
society." It was not simply problematic because "many things [children]
will need to know have not even been discovered yet," but also because
"any job" a child takes "will change, perhaps even disappear, within a few
years." He suggested that this was an era when "machines are cheap and

plentiful" and "human manpower is scarce and expensive." Therefore, "to manage our public affairs, we need a higher level of general education for all citizens than any previous society." Like many of the reformers of the period, Rosenbloom was quite optimistic about what children could manage at young ages—"children, if taught properly, can learn undreamed-of things"—and he encouraged teachers and administrators to work alongside scholars in order to develop a successful curriculum.[61]

Rosenbloom also warned that the failure to create a "mass movement" in support of new curricula risked "our very survival." People needed to act quickly because there was no easy mechanism for change and "neither the external enemy nor the social forces within will allow us much time."[62] Rosenbloom's words, in the context of a national conference on curriculum change, served as a reminder that reform was about more than updating textbooks or raising test scores. The new math was said to address the concerns of students facing the particular challenges of the postwar world; statements about the proper way to teach mathematics were embedded in conceptions about the current problems of social and political order.

Claims about the role of mental discipline for the challenges of the contemporary world took on a very specific valence within federal politics, as schools were increasingly configured as a crucial part of presidential proposals. In February 1962, President John F. Kennedy reminded Americans that although "education is both the foundation and the unifying force of our democratic way of life," "our educational system has failed to keep pace with the problems and needs of our complex technological society." Facing a "massive threat to freedom," Americans needed to take "urgent" action. Alongside new initiatives for classroom building and teacher salaries, Kennedy praised the "excellent but limited work" of the National Science Foundation curriculum programs and promised he would expand funding for new instructional materials in the next year's budget.[63]

Only a few years later, Kennedy's successor promoted the now much broader new math curriculum as confirmation of the transformative possibilities of the Great Society. In seeking an unprecedented expansion in education funding, President Lyndon Johnson echoed the rhetorical distinction between tradition and modernity used to promote the new math: "We must demand that our schools increase not only the quantity but the quality of America's education. For we recognize that nuclear age problems cannot be solved with horse-and-buggy learning."

Johnson claimed that the three Rs of education needed to be supported by three Ts: "teachers who are superior, techniques of instruction that are modern, and thinking about education which places it first in all our plans and hopes." As evidence, Johnson pointed to the "exciting experiments in education" already under way, including the fact that "many of our children have studied the 'new' math."[64] Johnson's comments were in support of what would become the expansive Elementary and Secondary Education Act of 1965, a crucial part of his belief that—regardless of past evidence—"the answer to all our national problems comes down to a single word: education."[65]

Kennedy's speech had been firmly in the original tradition of SMSG, promoting reform as a collaborative effort to address the rising demand for mathematically trained individuals in a competitive Cold War world. Kennedy's position on education was explicitly within the context of a "massive threat to freedom."[66] Indeed, *the Russians are coming was the story*" at conferences promoting curricular reform to teachers, as G. Baley Price would put it years later.[67] In contrast, Johnson implied that the new math was part of domestic reform. Johnson's reframing, especially given the absorption of the National Defense Education Act by the Elementary and Secondary Education Act, had symbolic significance: the new math's origins in a defense crisis of scientific manpower had been replaced by its dissemination as part of broader social reform in the 1960s.[68]

Kennedy and Johnson yoked their initiatives to the new math in part because the reforms played well on all political sides. The curriculum was widely understood to represent a positive step in the preparation of students for the challenges of contemporary America. Liberals appreciated the government's active involvement in improving education, while conservatives praised the way the federal programs sidestepped progressive educators in favor of scientists and mathematicians. The conservative head of education in California, Max Rafferty—no friend to progressives—wrote approvingly of his own state's adoption of new math textbooks on the basis that "the old, comfortable ways are no longer enough." Even John Miles of the Chamber of Commerce, vehemently opposed to most forms of federal education aid, agreed that efforts like SMSG should be expanded.[69]

The widespread appeal of learning "modern" mathematics also aligned with judgments that the problems of the contemporary world needed scientific solutions. Illinois senator Charles Percy explained that "we need the fullest possible development of the capacity to think, to reflect, to weigh and judge, to make choices among alternatives, and to foresee the consequences of these choices. This is the modern mind we need—the mind of the scientist, the key executive, the mathematician."[70] "Modern" minds ought to employ rational, mathematical, and structural approaches to a broad range of problems. It was in fact commonplace to note that the modern approach should be a mathematical approach. From the development of rational choice liberalism to the flourishing of structuralist ideas throughout academia, intellectuals in a range of disciplines argued that the complexity of the modern world required formal, structural, and particularly mathematical, approaches to understand it.[71] Mathematical ideas and methods were increasingly incorporated into the highest levels of government, including Johnson's own adoption of the "Planning, Programming, and Budgeting System" from the Department of Defense in order to "rationalize" the appropriations process. It was entirely fitting that the chair of the subcommittee on the "Place of Mathematics in a Changing Society" would be from RAND, a company steeped in the notion that rational, calculating minds were needed to face the challenges of the modern world. The new math was one of many initiatives and programs that originated at the conjunction of scientific and military interests but were later incorporated into the domestic sphere. Local and federal governments increasingly relied upon RAND and similar entities, claiming the authority of rational—mathematical—solutions to complex problems of all sorts.[72]

Popular magazines also promoted the idea that the intellectual habits of mathematicians would be critical to solving the nation's complicated problems. William Morrell of the NSF was quoted in *Newsweek* claiming that "there is an intellectual content to mathematics that the educated man should have and want to have if it is presented in an interesting and exciting way. Not 'Here is a problem, solve it.' But rather, 'Here is a situation, think about it.'"[73] *Time* magazine proclaimed in 1964 that "math is not only vital in a day of computers, automation, game theory, quality control and linear programming; it is now also a liberal art, a logic for solving social as well as scientific problems."[74] Thinking well meant thinking like a mathematician.

* * *

One paradox of the new math was that its most successful incarnation—the fundamental alteration of the secondary-school mathematics curriculum—was also invisible. Most memories of the books and indeed most contemporary discussion of the reforms involved the elementary schools and their changing textbooks. From the point of view of SMSG, this was doubly unfortunate in that the organization did not ever receive lasting credit for the monumental effort in the secondary schools but did face any criticism for the elementary school reforms made largely outside of its purview.

Although high school and elementary school reforms were experienced in very different contexts and through distinct mechanisms, *the* new math became a concept divorced from the myriad reformers' distinct initiatives and instead linked to the commercial texts rapidly appearing in elementary classrooms. In some respects, the elementary case is ancillary: the reforms were originally focused on college-capable students and geared to reform the secondary school curriculum. In other respects, however, the elementary textbooks formed the centerpiece of the new math, because it was through arithmetic that the claims for a new kind of mental discipline were presented most baldly. There wasn't anything wrong with the existing elementary textbooks per se, and supporters of the new math did not promise that the curriculum would foster improved computational abilities. Rather, reformers claimed the challenges of the contemporary world might be addressed, in part, by changing how first graders learned to add.

At both the primary and secondary levels, reformers succeeded in making their visions a reality. It took a great deal of effort and risk on the part of textbook writers, publishers, school boards, administrators, and teachers to change the curriculum so rapidly. The ultimate spread of the new math was evidence for the persuasiveness of the claim that mathematics made the modern, technological world, and that mathematical habits of thought were needed for solving the problems of this world.

Nevertheless, much had changed from the time of SMSG's first writing session in 1958. By 1965, the distance of education reform from any Cold War manpower crisis was palpable. The new math was sold as part of a social transformation, tied to Johnson's education legislation and to confidence in elite reformers and government programs to truly fashion a Great Society. As faith in modernism, federal interventions, and the transformative power of education faded by the early 1970s, so too would the new math's reputation diminish.

CHAPTER SIX

The Basic Subject

New Math and Its Discontents

Anyone who knows anything about the new math knows that it failed. Whatever qualifications may be placed on that assessment, the curriculum's fall from grace was undoubtedly rapid and decisive. What had been widely hailed as critical for the modern student in 1965 was roundly condemned a decade later.

This chapter explores critical responses to the new math, focusing on the early to mid-1970s. Reactions took many forms, from revised state curriculum guidelines to heated op-ed articles, from bold declarations of a return to "basics" to quiet revisions of classroom practices. Although some critics pointed to lower test scores and the presumptuousness of elite reformers as evidence for and causes of the curriculum's failure, the main complaint of the new math's detractors was that the curriculum instilled the wrong mental habits. They claimed intellectual training should entail basic skills and rote exercises, not structural understanding or abstract reasoning. The critics of the 1970s did not simply oppose the pedagogical or mathematical approach of the new curriculum as had some mathematicians and teachers in the 1960s, but rather challenged the very conception that an intelligent citizen ought to understand the structure of mathematical knowledge.

The fall of the new math, like its rise, was thoroughly political. That is, one way of reading the trajectory of the new math is to understand it as inseparable from the broader political ferment of the period. Critics of the new math often explicitly aligned themselves as part of the conservative revival and "right turn" in American politics, and their understanding of the math curriculum revolved around questions of moral authority and expertise. Their views of the intellectual habits math should provide contradicted those of the politicians, educators, and professional mathematicians responsible for the new math. The curriculum was ultimately as political in its reception as it had been in its promotion. Debates about it remained debates about mental discipline and social order.

The Demise of the New Math

According to Gallup pollsters, the new math was by far the most widely recognized educational innovation of the postwar era. Howard Fehr, one of the leaders of mathematics education in the period, estimated that by 1965 nearly 75 percent of high school students and 40 percent of elementary school students nationwide were studying a version of "modern" mathematics; a decade later another survey estimated some form of the new math was present in 85 percent of schools. When Julie Nixon Eisenhower was training to be a teacher, it was newsworthy—but not surprising—that she took a "new math course" as part of her training.[1]

Despite the fanfare accompanying the increasing use and visibility of new mathematics textbooks in the middle part of the 1960s, the nation's attention rapidly moved elsewhere. Teachers adjusted to the new materials, schools formalized the revised curricula, and parents quietly struggled to keep up with the new questions their children brought home from school. One reason for this silence was that the new math was never as extensive as Fehr and other fervent promoters (and critics) believed. The head of one new math program, the Madison Project, complained that even at the height of popular discussion of the new math, many schools either ignored the reforms or used textbooks that included only the most superficial of changes. At best, Robert Davis claimed, a "significant minority" of schools were doing something that reformers would actually consider an improvement.[2] Many districts remained poor, relying on decade-old books in order to save money. Furthermore, rural schools with limited enrollment could not consistently offer a full range of traditional mathematics courses, let alone redesign existing courses or add new ones. One survey of Missouri schools in 1966–67 found that secondary schools in that state offered a course in mathematics only on alternating years in order to be able to offer upper-level mathematics at all. Missouri was not alone. Many high schools continued to require only one math course for graduation; the national mean requirements for math education essentially held steady from 1958 to 1972; and the percentage of students taking math courses remained constant over the 1960s.[3] The spread of the new math was uneven at best.

The complaints that did emerge in the 1960s primarily focused not on the curriculum's value but on adequate preparation for it. Parents certainly complained about the unfamiliar math problems their children brought home. Individual schools and the National Science Foundation worked to educate parents on the reasons behind the changes and provided opportunities for parents to learn the material.[4] By the turn of the decade, though, the second generation of new math parents received little or no training on the curriculum and instead came face to face with the new texts only when helping their children with homework. Even in Winnetka, Illinois, a community on the forefront of curriculum reform (the district had enrichment classes on the new math for parents as early as 1962), parents' knowledge of the new math correlated with approval of the curriculum, and rates of familiarity were dropping by the early 1970s. The new books indeed looked strange—and certainly aroused suspicion among parents convinced that math should not have changed that much from when they learned it. Similar claims also emerged from elementary teachers, who, like parents, were not subject experts in mathematics. Their opinion about how math should be taught was easily overruled—and, in the case of most curriculum development organizations, not even consulted. Alongside some mathematicians' grievances about the portrayal of their field, early critics' primary complaint was that the introduction of new textbooks into elementary schools went "too far too fast."[5]

Even facing this criticism, the new math's demise was hardly inevitable. SMSG's own end was brought about by financial pressure from above, not discontent from below. After 1969, in the midst of new National Science Foundation budgetary pressures created in part by the war in Vietnam, SMSG's budget was halved—its request for $1.2 million reduced to just $641,000.[6] These budget cuts eventually forced Edward Begle to concede that the mission of SMSG could be effectively dispersed to other forums, and he announced that SMSG would be ending its formal operations in the summer of 1972. At that point, however, its leaders found little reason to be apprehensive about the long-term success of the program, and, in initial post-mortem analyses, the consensus among commentators was that SMSG's efforts had served to raise the profile and quality of math in schools across the nation.[7]

It did not take long, however, for the new math's reputation to decline precipitously. In 1974, Ohio representative John Ashbrook decreed the new math "passé" and lamented the federal government's education programs as failed experiments that had promoted only academic "frills." Ashbrook, who had run against President Richard Nixon in the 1972 primaries as a "principled conservative," now asked, "How much in Federal dollars went to further this innovation? Also, how many other, less well known, 'innovations' have actually hurt schoolchildren?" Ashbrook's concerns were not limited to the far right. By late autumn 1973, just over a year after the largely positive retrospectives of SMSG, a *New York Times* article quoted mathematician Morris Kline's description of the new math as an "irresponsible innovation" and portrayed the curriculum as widely criticized by mathematicians, teachers, and elected officials.[8] These criticisms were focused not on problems with teacher preparation, but on the failure of the reforms' basic premise.

Between the end of SMSG and Ashbrook's comments, a series of well-publicized complaints documented and perpetuated a shift in public opinion against the reforms. A front-page article in the *Los Angeles Times* covered the crusade of a state assemblyman—and chair of the California Assembly Education Committee—against the new math. Leroy Greene, an engineer by training, claimed that when he looked at his daughter's elementary school new math textbook, he could tell that computational skills had been neglected at the expense of theoretical concepts and problems. Greene called a public hearing in the spring of 1973 and convinced the legislature to require the California Board of Education to adopt a new series of books that emphasized computation.[9] In May 1973, the *Chicago Tribune* reported that in one Chicago suburb using new math textbooks, the eighth grade math scores had been on the decline for three consecutive years. In New York State, the Fleischmann Commission, primarily concerned with the institutional and financial support of the schools and teachers, noted that too many students were doing poorly in mathematics. New York's chief of the Bureau of Mathematics Education, claiming to be acting on the behest of teachers and with the backing of the Fleischmann Commission's evidence, said in January 1973 that the state curriculum guidelines would "swing back" away from sets to an emphasis on computational skills. During the same school year, state officials in New Hampshire asked schools to emphasize computation to make up for the failings of the new math textbooks.[10]

In 1973, Morris Kline's *Why Johnny Can't Add* appeared, lending the support of a professional mathematician to the claim that the reforms were mathematically and pedagogically flawed.[11] While Kline had been complaining of the new math for well over a decade, his previous readership had largely been limited to curriculum designers and mathematicians. Kline's criticism now found a much larger audience, and the book was widely reviewed. (Tellingly, a similar diatribe against the new math published just a few years before Kline's was almost completely ignored.)[12]

When newspaper articles or critics did present specific evidence against the new math, it almost always consisted of reports of declining computational ability, as measured by standardized tests. California state math scores were said to have dropped in the three years after 1968, and the *Fleischmann Report* had claimed that nearly a third of New York's sixth graders were performing unsatisfactorily in mathematics, as measured by computation tests. In New Hampshire, hard numbers were rarely cited, but the state's math consultant claimed that "since the advent of modern math, the computation scores of students in the state have very definitely declined." The experiences of these states, along with later data of depressed SAT scores, were repeatedly hailed as decisive indicators of the corrosive effect of the new math on student learning.[13]

From the outset, the use of test scores as evidence against the new math raised more questions than it answered. In California, test scores in *all* school subjects—as well as IQ scores—were decreasing in this period, and general declines were occurring across the country as enrollments increased throughout the 1960s.[14] The *Fleischmann Report* in New York was not in fact uniformly negative: it documented both increases and decreases in elementary mathematics scores.[15] When set alongside control groups of older Americans, students who had studied the new math were found to have computational abilities that were comparable to their elders—if not better.[16] The data for a prominent front-page article in the *Wall Street Journal* were later questioned on the basis that the claimed score "decline" was possibly an artifact of testing conditions.[17]

The link between the new math and declining SAT scores was just as suspicious. The slump in math scores was, in fact, less than that for verbal scores from 1962 to 1975. Two issues of the journal *Educational Technology* were devoted to explaining the test score decline and concluded that the greatest change was in high school students' scores, mainly involving verbal skills, and was most pronounced in the population of female test-takers. The authors surmised that perhaps as more women planned to attend college over the course of the 1960s and 1970s, a broader population was taking the tests and depressing the average. Other than demographic explanations, they simply suggested that the decline might be best explained by changes in the courses students were taking (i.e., not in the quality of education they received per se) and the relationship of that material to the tests.[18]

In some cases, even the very fact of declining scores was contested. Many of the cited tests tracked scores of a single age group, rather than the computational ability of individuals over time. Defenders of the new math claimed that the novel curriculum might reduce computational ability in the lower grades but eventually lead to higher scores in upper grades. SMSG's own studies showed that there was as much score variability among students using different new math textbooks as there was between students using either new math or traditional textbooks.[19] For every study or article that showed a decline, there was another indicating that certain new textbooks might actually improve scores, or at least stabilize previously falling scores. One report suggested that in the midst of general test score declines, those in new math courses actually fared better; another argued that new math was the best preparation for college math courses.[20]

In any case, all the claims of decline were based on the presumption that previous baselines existed: the fact was that the new math had ushered in the era of curriculum efficacy measurement, so there was little to compare the reforms against. Begle and his associate Paul Rosenbloom had introduced the idea of rigorous testing and analysis of the *comparative* and *relative* effects of math curricula on student learning. They not only tested their own books but also created the first National Longitudinal Study of Mathematical Ability. Prior to the new math, extensive testing of curricula was essentially nonexistent, further challenging any sweeping claims about the effect of the new textbooks on test scores.[21]

More significantly, testing of computational ability missed the entire point of the reforms. Begle knew from the outset that computational ability might decline somewhat and was willing to accept minor declines in exchange for greater conceptual understanding and facility with mathematical thinking. One SMSG member of the ninth grade writing team explained, "skills *are* important, but never as ends in themselves."[22] As another SMSG textbook writer—and high school math teacher—noted when promoting the new secondary school curriculum: "If my students taking the new program can hold their own with traditionally taught students on the traditional tests in the early stages, I am satisfied."[23] Begle suggested that if schools wanted to artificially raise computational scores, teachers should just spend a few minutes drilling students on arithmetic in addition to teaching out of the new textbooks—but ultimately, rote computational ability should not be the goal of mathematics education.[24]

The claim that the new math had hurt students' mathematical ability, in short, was highly suspect, at least on the basis of test scores. The change in public opinion, however, was very real. Popular opinion of the new math would never recover after the negative media coverage of the early 1970s. Ashbrook was not railing against a popular program in 1974 but rather was simply one of many voices turning against the curriculum.

Behind the façade of falling test scores, three interrelated factors were responsible for the proliferation of complaints about the new math. First, criticism—especially of elementary school reforms—was increasingly embedded in a broader critique of federal programs, and critics of the new math often doubted the ability of elite, national activists and curriculum writers to properly shape the intellectual training of American students. Such doubts were particularly driven by a diminishing faith in the efficacy of top-down reforms in the years following 1965. Second, the context of curricular reform had been transformed. That is, reformers looked to the schools for very different reasons in the late 1960s and early 1970s than they had a decade earlier and found that the original basis for updating the curriculum was no longer compelling. Finally, and most important, those opposed to the new math began to challenge SMSG's claims about the desirability of teaching "modern" mathematics at all. Critics pointed to the need to raise computational test scores but also to the need to cultivate traditional habits of thought in a world that seemed to have lost its moral grounding. The criticism accused the new math not just of being emblematic of federal overreach during the liberal 1960s (however historically misconstrued that might have been) but also of providing students the wrong sort of intellectual training.

Expertise and the State

When Seymour Sarason, director of Yale's Psycho-Educational Clinic, set out to examine theories of reform within school cultures in the late

1960s, he took as his central case study the introduction of SMSG's new math curriculum into an elementary school. Sarason discovered that while the individual agency of teachers and administrators affected how curricular changes proceeded, the structure of educational institutions was far more important. In particular, he noted that strongly held local theories about teaching proved to crucially undermine the intentions of elite—and distant—curriculum writers. The role of teachers, he suggested, was in fact "antithetical" to being a leader or vehicle of change.[25] SMSG had made teacher commentaries as thorough as possible—trying to counter teachers' lack of knowledge or willingness to experiment—but ultimately the gulf between textbook writers' intentions and the traditions of teaching could not be bridged. The new math entered the elementary schools largely "from above," creating an obstacle to its acceptance.

On the basis of his experience watching schools incorporate the new math, Sarason concluded that curricular reforms would be resisted explicitly or implicitly by the culture of elementary schools. His finding was supported by near-contemporary studies that found teachers' length of tenure had a strong negative correlation with approval of the curriculum. It seemed that more experience meant more resistance to change. And years later, National Science Foundation interviewers would agree with Sarason's conclusion that teachers resisted curricular innovations because they challenged the teachers' competence and voice in the educational process.[26]

The new math's dissemination occurred against a backdrop of changing student demographics, as well as changing attitudes toward educational interventions. After 1965, with the slowing of the baby boom, enrollments declined, and schools ceased to be considered the panacea they had once been. Both the 1965 Moynihan Report and 1967 Kerner Commission attempted to explain the persistence of social strife, particularly racial inequalities, by pointing to the inability of federal action to solve social problems. Further evidence came from the 1966 Coleman Report, which suggested that socioeconomic levels, values, and family culture served as better indicators of future academic achievement than the type of school attended or amount of money spent on the student in school. Later research by Christopher Jencks confirmed that spending money on education did not seem to lead to desired "social outcomes," perhaps owing to the impossibility of disentangling education from other factors. The result of this cluster of reports was a growing sense that the problems of society would not be solved simply by promoting the proper forms of mental discipline—*Time*'s review of Jencks's work was titled "What the Schools Cannot Do."[27] Thinking well—thinking mathematically—might have benefits, but the optimistic pronouncements of Begle and other officials that changing textbooks would have a meaningful effect on social outcomes was increasingly suspect.

Moreover, faith in academic or elite forms of knowledge suffered from a general decline in the perceived authority of universities and of the "expert" knowledge such institutions produced. The new math was predicated on the assumption that the intellectual habits of academic mathematicians would be good to cultivate in students. By the 1970s, however, it was not clear that academic knowledge might actually be beneficial for the rest of society. Colleges came to be associated with protests and radicalism, and with being increasingly isolated from "mainstream" America. By the 1970s, professors no longer enjoyed as much cultural prestige, and fewer Americans looked to "the best and the brightest" as sources of reliable practices or authoritative knowledge claims.[28]

Along with declining faith in universities as authoritative cultural institutions and in schools as agents of social change, approval ratings of the federal government itself plummeted. Budgetary pressures and the cultural divisiveness of the Vietnam War had weakened optimism about President Lyndon Johnson's Great Society. Labor unrest peaked in 1973–74 as government and industry seemed less and less able to protect the livelihoods of American workers.[29] The slowly unfolding Watergate scandal, the resignation of Richard Nixon, and the ignominious withdrawal of troops from Vietnam all helped to erode confidence in governmental institutions—and these events occurred simultaneously with the emergence of news stories critical of the new math. Tellingly, a May 10, 1974, front-page article in the *Los Angeles Times* announcing California educators' increasing skepticism toward the new math directly followed a lead story about the opening of the Nixon impeachment hearings.[30] Perhaps more damning than general approval ratings was the perceived failure of government interventions. Daniel Patrick Moynihan, a one-time architect of social programs, lamented that federal initiatives, particularly those intended to alleviate poverty, had backfired. He warned in 1967 that there was "a certain coming together in opposition to, in distaste for, bigness, impersonality, bureaucratized benevolence and prescribed surveillance . . . it is the vibrant, established, *coming* young people of the nation who in large numbers have learned to distrust their government,

and in many ways to loathe their society. . . . their understanding of
their country will have been shaped by the traumas of the 1960's."[31]

Moynihan's prediction was generally accurate: faith in government
declined rapidly. The percentage of citizens who trusted the federal government "most of the time" or "just about always" dropped from 76 percent in the mid-1960s to 33 percent in the mid-1970s and 25 percent by
the end of the decade.[32] And as domestic disillusionment grew, so theories based on the supremacy of the "modern" American ideal came to
seem far less persuasive.[33] The shift was never complete, of course, and
neither American faith in government nor the international standing of
the United States ever irreparably plummeted. Nevertheless, if American society in the 1960s represented the high point of modernity talk,
curricula premised on preparing "modern" students soon became a hard
sell indeed.

While the two sources—government and academia—whose collaboration enabled the new math faced new crises of confidence in the 1970s,
perhaps most surprising was the decline in support for the National Science Foundation and its programs. In the late 1950s and early 1960s,
the NSF was one of the most trusted of government agencies, and the
emergence of SMSG and other curriculum groups was due in part to the
sense that programs led by scientists rather than educators avoided any
unseemly display of federal influence on education. The trust placed
in scientists—who often claimed that what they did was expensive and
opaque but required for the well-being of society—declined as the Cold
War wore on. Cultural and political criticism of science and technology
became widespread, from the increasing skepticism of the safety of nuclear power to the mockery of Mutually Assured Destruction in Stanley Kubrick's film *Dr. Strangelove* (1964). Rachel Carson's 1962 *Silent
Spring* exposed a generation of readers to the possibility that "better
living" through chemical pesticide use had a significant and dangerous
dark side. In constant dollars, defense funding for science decreased in
the early 1970s, and Nixon disbanded the President's Science Advisory
Committee (PSAC) soon after his reelection in 1972. Scientists themselves, even those who once had benefited greatly from military funding,
began to distance themselves from the Department of Defense. Despite
the popularity of the space program, Americans were increasingly wary
of the underlying rationality and benevolence of the powerful military-
scientific nexus.[34] Under greater scrutiny, NSF projects, particularly the

infamous Mohole fiasco, a grandiose and ultimately abortive attempt to drill down to the Earth's mantle, appeared to be expensive frivolity at best and wasteful nonsense at worst.[35]

Most damning, perhaps, to the trust placed in scientists for the design of curricula was the controversy over "Man: A Course of Study." Meant to be an introduction to human communities around the globe, this social science curriculum faced a small but vocal group of congressional critics who charged that it spread a harmful version of cultural relativism and presented explicit descriptions of disturbing practices to children. Representative John Conlan of Arizona, one of the harshest critics of the curriculum, retrospectively linked this "blunder" to the new math.[36] By the 1970s, it was nearly impossible to consider the new math separate from these other controversies. The leeway given mathematicians and scientists to promote strange or unfamiliar practices in precollegiate classrooms had disappeared. "Man: A Course of Study" seemed to prove that what was done in universities had no place in the high school or elementary school curriculum. While the new math's productive period was long past by the time of this particular controversy—and so the only effect was on evaluation of the new math—the uproar led directly to greater congressional oversight as well as to the formal end of many NSF curriculum development projects in 1975.[37]

The Shifting Landscape of Reform

As the 1960s wore on, changes in the connotation of educational reform also affected the new math's reputation. The desire of educational reformers in the late 1950s to provide better intellectual training eventually gave way to a focus on the inadequacies of inner city schools, the persistence of American poverty, and the challenges of educating culturally "disadvantaged" students. Indeed, much of the criticism of the new math was grounded in an assumption that schools ought to address problems of inequity by ensuring that all students receive at least some minimal level of education. Just eight years after the *New York Times* had run a front-page article heralding the arrival of the new math in 1965, the same paper described the "counter-revolution." With the race to the moon successfully won and post-Sputnik fears abating, the concern of educational reformers shifted to the slums and ghettos, where students did not display even basic proficiency in mathematics.[38] Organizations like SMSG, which hoped to teach *all* students essentially the *same* sort of mathematics, seemed increasingly naïve.

Harvard psychologist Jerome Bruner, in a 1971 retrospective of his 1960 book *The Process of Education*, described this shift in the dominant stance of education reform. His book had provided one of the most important psychological justifications for the new math; Bruner believed students were able to learn more "modern" mathematics than traditionally assumed, and that learning the structure of mathematics would facilitate the development of intellectual discipline generally. Looking back, Bruner described how his optimism in this process was a matter of "faith": designers of the new math assumed that students had absorbed the "middle-class hidden curricula" and wanted to develop the "traditionally intellectual use of the mind." By the 1970s, he realized that reformers' failure to respond to "social needs" had pushed schools further into "crisis" mode. Students in many settings simply did not have the motivation to develop their intellects. Debates about "modern" math and intellectual discipline seemed ridiculous when some students lacked even basic access to quality schools and teachers. Bruner concluded that reformers now had to consider fundamentally "refitting" schools for the "needs of society"—which would require reforming the institution, not just changing the curriculum.[39]

A new critical genre had indeed emerged in the late 1960s, confirming that educational reform increasingly meant institutional reform. These critics focused on the infrastructure, routines, and dominant ideologies within schools. In 1967–68 alone, Jonathan Kozol criticized the impoverishment of militaristic urban schools in *Death at an Early Age* (1967); Philip W. Jackson analyzed the routines and rhythms of the elementary school in *Life in Classrooms* (1968); Louis Smith and William Geoffrey observed an urban classroom and attempted to provide a theoretical framework for it in *The Complexities of an Urban Classroom* (1968); and director Frederick Wiseman released *High School* (1968), a film exploring the values imparted by the daily grind of a Philadelphia school.[40] One impetus for these projects was the growing attention academic philosophers and sociologists paid to the institutional role of the school. Erving Goffman's examination of "total institutions," Basil Bernstein's elucidation of class "codes," and Pierre Bourdieu's studies of "reproduction" all spoke to the emerging significance of placing the school and curriculum within a matrix of discipline and power.[41] These themes were not entirely new, of course, and the social purposes of the school had long been contested, but the rapid emergence of so many critiques suggests a shift in the way educational reformers conceptualized the problems within schools. Far from a neutral vehicle for the dissemination of curricula, the school itself mattered as a social institution. Schools were more than the sum of their curricular parts.

The debate over whether the federal government should have a role in education yielded to a debate over how it should intervene. The Elementary and Secondary Education Act of 1965 had established a permanent federal role in the schools, and most discussion of the new math in Congress in the 1970s was in reference to this new reality.[42] That is, the switch of new math's context from the National Defense Education Act to the Elementary and Secondary Education Act did more than shift discussion from college-bound students to elementary ones. It meant that the curriculum was evaluated in the context of domestic reform programs, not international, scientific manpower initiatives.

In this new context, there was a common perception that schools had not only been unsuccessful in reducing poverty but were also failing to provide the poor with an adequate education. Progressives tended to emphasize the need for more flexible schools and open classrooms, making education less militaristic, while conservatives favored a push for accountability, especially emphasizing test-based measurement. "Accountability" and "standards" became widespread within educators' lexicon in the 1970s, as educators and state officials attempted to ensure that the schools were adequately serving their populations.[43] In the push to hold schools responsible, states began to pass legislation promising "education accountability," making it state law that individual schools meet minimum performance standards.[44] The development of the National Institute of Education in the 1970s was in part meant to replace the traditional statistical function of the old Office of Education with a research-based organization that promoted the study of school accountability.[45] While Begle had believed that all students could master mathematics using SMSG's books, SMSG's writers never thought their fundamental purpose was to ensure that every student could pass a basic test of arithmetic. SMSG's concern was with the proper development of individual minds, not with fundamental institutional reform or with meeting minimum standards for a particular socioeconomic subgroup.

If the new math fell through cracks in the shifting ground of educational reform, it was also a victim of infighting among the bureaucratic organizations responsible for its funding and oversight. The President's Science Advisory Committee, which had been so influential in convincing Eisenhower to support the initial NSF curriculum projects and teacher training institutes, gradually shifted over the course of the 1960s to a focus on doubling the number of doctorates in scientific fields. Even before its elimination by Nixon, the committee had moved decisively away from precollegiate concerns.[46] While Congress continued to support the NSF's educational work, particularly the summer institutes, the NSF itself promoted university research funding above all. An internal organizational change in 1965 divided the NSF's educational division into separate precollege, undergraduate, and graduate divisions, effectively giving greater emphasis to graduate education. Finally, the Office of Management and Budget was ever more skeptical of general aid to education after 1970 and consistently cut NSF funding for broad education programs in favor of specific projects for schools and districts.[47] The NSF's Course Content Improvement Program was too expensive—and the benefits too diffuse—to justify its cost in the budget-cutting years of the early 1970s. Among both educational reformers and bureaucratic authorities, the new math simply lost its constituency.

The Politics of "Back to Basics"

The diminishing authority of the federal government and professional scientists and the changing nature of educational reform provided important contexts for understanding why criticism of the new math proved so effective. The criticism itself, however, focused directly on the politics of knowledge. Critics rejected claims made about the mental discipline that new math was said to promote. Students facing the challenges of the contemporary world needed the intellectual habits provided by traditional mathematical knowledge, they argued, not those provided by "modern" mathematics.

The stakes of this distinction were made explicit by a highly publicized *Washington Post* article about a parent's experience with the new math. In fall 1972, James M. Shackelford had opened the pages of his daughter's elementary math textbook only to find problems that he himself could not solve. Noting that the questions involved somewhat abstract ideas about sets, he was disappointed to find that the textbooks required much less old-fashioned memorization. (Shackelford's daughter was using the popular Houghton Mifflin series Modern School Mathematics—Structure and Use, partially modeled after SMSG's materials.) Later, he asked his daughter and her friends to calculate 8×9 and was shocked to find that none of these fourth graders could do so. Despite being a Ph.D.-holding chemist employed by the U.S. Environmental Protection Agency, he claimed that the most abstract math he used was that needed to figure out the grocery bill. Shackelford's article condemning the new math led to dozens of heated letters to the editor. The published letters were evenly split on Shackelford's crusade, but the sheer volume suggested the issue was volatile; one writer noted that more letters had been printed on the topic of the new math controversy than on that year's presidential election.[48]

Shackelford's critique did not include any evidence beyond the informal survey of a handful of fourth graders. The article's resonance, however, emerged from Shackelford's suggestion that the mathematical knowledge required to face the problems of contemporary life mainly involved memorizing the multiplication table. In contrast to SMSG's emphasis on the creative, flexible, structural thinker, able to reliably move from postulates to sure conclusions, Shackelford and his supporters conceived mathematics as the means by which fundamental skills were established, traditions were renewed, and memory exercised. That is, they asserted that citizens should receive training in mechanical skills and memorization techniques, rather than be given a conceptual or structural understanding of the nature of mathematical knowledge. Complaints like Shackelford's would help usher in a movement eventually known as "back to basics."

The sensibility represented by the "back to basics" cause epitomized the backlash against the new math in the 1970s. Proponents of "back to basics" claimed that an overemphasis on mathematical structure in the elementary schools had impeded the ability of students to calculate, and they encouraged schools to return to the "basics" of drilled exercises and rote memorization of arithmetic facts. A large section of the March 1977 issue of the journal for teachers, *Phi Delta Kappan*, was devoted to the "back to basics" movement, which its editors noted was "today's media event." The Gallup organization recorded consistent support for—and awareness of—the "back to basics" movement throughout the mid-1970s. One contemporary newspaper article called it the "most-talked-about" movement in education. Recognizing this new public infatuation with the "basics"—and not wanting to appear opposed to competence in com-

putational skills—the National Council of Teachers of Mathematics devoted a February 1978 article in the *Mathematics Teacher* to explaining
how the organization and its allies indeed supported "universal competence in basic computational skills," but as part of a "total mathematics
program" that focused on mathematical understanding alongside other
essential mathematical skills.[49]

"Back to basics" was overwhelmingly a collection of local, decentralized efforts. There was no overarching organization; even the pointedly
named Council for Basic Education had been formed in the 1950s to
promote academic disciplines generally (not rote learning) and was certainly not uniformly supportive of the new movement.[50] Instead, critics
usually took action at the level of individual classrooms, schools, and districts, citing mainly their own experiences but also recycling complaints
like Shackelford's and the arguments of Morris Kline's *Why Johnny
Can't Add*. In some states, parental discontent combined with reports
of declining test scores—however spurious their direct connection with
the new math—to convince state boards of education to recommend the
replacement of textbooks. The combination of a decentralized system
of education and a lack of coordination ensured the movement's results
were uneven and locally dependent.

"Back to basics" critics of the new math echoed its proponents in emphasizing a clear link between instruction in mathematics and intellectual discipline. This was evident in media coverage. In one article, the
reporter combined a parent's complaint about the curriculum—"They
have gone overboard on teaching sets and what they call the new math.
I don't think any kid through third grade can multiply 3 times 3 and
get 9"—with an analysis of the consequent decline in behavioral habits generally—there was "total disorganization, rowdyness, children allowed to slam books around. I can remember when I was a child we were
taught basic respect, manners."[51] A return to drills in arithmetic entailed
a return of discipline. In Pasadena, California, a new school called itself
the John Marshall Fundamental School and advertised homework starting in kindergarten, daily character-building discussions, a "behavioral
adjustment room," and—explicitly—no new math. A feature article in
Newsweek in October 1974 tracked this new fundamental school in comparison with the "progressive" Pasadena Alternative School, portraying
the "fundamental" school as the new trend in education.[52] The reigning
educational movement in the 1970s was toward eliminating "frills" in an
effort to promote memorization and rote discipline.

The "back to basics" movement gave a new twist to the old idea of basic education. Critics like Bestor in the 1950s had pushed basic education in the sense of emphasizing the traditional liberal arts. By the 1970s, basic education proponents focused instead on rudimentary literacy and numeracy. A focus on computation does not necessarily imply a desire for mindlessness; historically, reckoning ability could count as evidence of genius.[53] Emphasizing the basics in the 1970s, however, was not at all about clever symbolic manipulation or extraordinary mental reckoning. It was simply a matter of emphasizing the memorization required for "competency" in calculation. This shift in what counted as "essential" inverted the intellectual virtues of the curriculum. The new math had once been promoted as the pinnacle of rigor, introduced by research mathematicians in order to shore up the nation's flagging intellectual discipline. Opponents in the 1970s believed that the nation needed to go back to basics for precisely the same reason—the promotion of mathematical rigor.

Teachers themselves pointed to intellectual virtues when discussing the benefits of traditional mathematics education. In the mid-1970s, the National Science Foundation conducted extensive case studies, surveys, and literature reviews as part of its effort to evaluate the fate of the course curriculum improvement project. One component of the studies included NSF researchers interviewing teachers and administrators in eleven school districts scattered across the country. The interviewers overwhelmingly found that teachers rejected the justifications for the new math in favor of a return to traditional skills. An elementary teacher explained that "we are terribly old fashioned and I am proud of it. It is old fashioned and super to expect every first grader to have 'rapid memory' of basic facts to ten. . . . When it gets down to it every teaching technique that works is an old fashioned one that involves understanding facts and remembering them quickly. The really able children and people are the ones who have the concepts under the rapid memory." Another high school teacher across the country concurred: "We've found that traditional methods work. This is the way it was taught to us in high school and the way it was taught in college and the way it works for us." Many of the interviewed teachers proclaimed themselves to be "very traditional." The old method of teaching cultivated the discipline necessary for the good of society: "I still think America came farther and faster than any nation in history under the old method of teaching, where we had some discipline in the classroom, we did some drill because it was

what teachers deemed was necessary, [and] we didn't have to try to justify all we did."[54] Discipline, derived from drills, drove the nation's development. Times were tough now, the teachers seemed to imply, precisely because we had moved away from "old methods" and traditions.

The interviewed teachers also explicitly dismissed the new math. One
elementary teacher in Texas recalled that "six years ago we made a mistake and went too far in teaching abstractions. We had to; we had no
choice. All the texts were modern in Texas." This was a mistake in large
part because "there is no way on God's green earth to teach mathematics without a lot of drill." A new teacher noted that "in my mind modern mathematics was an unfortunate hoax." A more experienced one explained that "I dislike our book, not enough drill, it's modern math. We
adopted a new book. I don't know its title but it has more drill, more
basics, and I'll like it." [55] This particular teacher did not even bother to
learn the name of the new book but knew that it included more drills
and was, therefore, better. The point is not that these opinions were universally shared but that it was commonplace for teachers to talk about
the virtues and vices of particular math curricula in terms of intellectual training, discipline, and social order. "Back to basics" was undoubtedly driven in part by a fear of declining test scores, but these teachers
seemed to confirm it was about a lot more.

It is profoundly ironic that an essentially conservative midcentury
Cold War project—one funded by the National Science Foundation, led
by professional mathematicians, and intended to reinstate mental discipline in schools supposedly softened by the reign of progressive educators—would become a target of conservatives less than two decades later.
The new math had come to seem like another educational experiment
gone awry. Teachers and parents lumped the new math with other innovations of the 1960s, from open classrooms to "audio-visual" education,
and saw them all as failing to address the challenges of preparing modern citizens.[56] As a *Newsweek* article concluded, "On balance, then, the
growing call for a return to the basics seems a healthy signal that masses
of Americans are no longer willing to accept at face value the pharma-
copoeia of educational nostrums that has been handed them by a relative handful of well-meaning, but sometimes misdirected innovators."[57]
One 1973 reviewer of *Why Johnny Can't Add* concisely captured this re-
visionism when he listed the new math as just another product of the previous decade's excesses: "It seems that one of the major concerns of the
1970s will be the dismantling of the great structures that we erected with

such pride in the 1960s. Project Apollo is gone, the Great Society is going fast and Vietnam is dying a lingering death. Morris Kline, a distinguished mathematician, now says that the new mathematics should be added to the list."[58] The specific intellectual justifications for the new math were ignored in this newly configured historical narrative, a narrative bolstered by Johnson's own hijacking of the new math as part of the Great Society. By the mid-1970s, the new math could be dismissed as just another failed federal program of the 1960s.

Although the "basics" movement was diffuse, political operatives quickly realized the potential to harness discontent at the local level for national causes. Conservatives recognized schools not just as venues for reinstating "traditional" values but also as places where parents could be recruited to a cause or mobilized politically. In the wake of Ronald Reagan's election, conservative activist Burton Yale Pines published a book that looked back to the 1970s to explain the transformations that had put conservatives into power. He believed the conservative triumph was simply the "back to basics" movement writ large. For Pines, the kaffeeklatsches of California parents promoting "basic education" candidates in school board elections were a part of the broader groundswell of "grass-roots" support for Reagan. Pines highlighted the manual "How to Win a School Board Election," written by a Christian fundamentalist, to indicate the degree to which mobilization on a national level often began with local elections.[59] Indeed, historians Lisa McGirr and Michelle Nickerson found similar midcentury avenues of conservative mobilization, as state and national politicians leveraged "local" issues. And when sociologist Sara Diamond looked for the roots of right-wing Christian politics, she also noted the importance of schools. Schools were not just places where "values" were perceived to be most under siege but also venues where relatively small groups could effect meaningful change.[60]

Conservatives were able to quickly capitalize on the "back to basics" movement in the 1970s in no small part because they had years of experience using local school concerns as causes célèbres. "Back to basics" supporters were drawing on established structures of opposition to what was portrayed as a "liberal" encroachment on local schools. Perhaps the most well known of conservative curricular activists, Norma and Mel Gabler, incorporated the removal of new math from Texas classrooms into their broader campaign against the adoption of "subversive" and "un-American" textbooks. Given the importance of Texas as a national textbook market and the vehemence of their rhetoric, the Gablers influenced the content of textbooks far beyond their state's borders.[61] Although not all critics of the new math counted themselves as political conservatives, debates over the introduction of sex education, the reform of history and civics textbooks, and the inclusion of new authors in literary anthologies in the late 1960s provided a template for those who opposed the new math. Usually conceptualized as elements of the "culture wars" and the rise of the Christian Right, these earlier debates were not just about the inclusion of specific beliefs or viewpoints in textbooks. Like contemporary skirmishes over prayer in schools, they were also expressions of concern about changing sources of moral authority. Activists in these earlier battles often saw themselves as fighting against programs designed by so-called experts and specialists who disregarded their "traditional" moral values.[62]

A similar charge would eventually be levied against the new math, as "back to basics" proponents outlined the importance of replacing the novel textbooks with ones emphasizing facts and memorization. The new math controversy was distinctive in the degree to which this argument was taken: parents and other laymen claimed they, not professional mathematicians, should be the ones to determine the content of math textbooks and mathematics' relationship with desirable habits of thought. Furthermore, the new math was the only reform effort that rooted federal money directly and explicitly in claims about how students should learn to think. The new math controversy uniquely *forced* districts to take a stand. The controversies over prayer and sex education in the schools may have been even fiercer, had so many concerned parents not taken their children out of the public schools entirely.[63] Sex education could be avoided, but *every* student learned math.

Many left-leaning commentators certainly saw "back to basics" as part of a broader conservative movement. A National Education Association position paper of 1978 noted that after military campaigns like Vietnam, it was not surprising to see the reemergence of neoconservative approaches, and to emphasize the good old days and the need for accountability in the present. The editorial page of the teachers' professional journal *Phi Delta Kappan* called the movement "nostalgia's child." Similarly, an article by the journalist Fred Hechinger in one of the official organs of the National Education Association, *Today's Education*, made the point that "back to basics" arose from "the distaste with which many educational as well as political conservatives have viewed the radical and neoprogressive currents of the rebellious sixties." Lamenting the return of autocratic discipline and corporal punishment, Hechinger also warned of educational conservatives joining forces with politicians who wanted to look tough on budgets.[64] Hechinger saw the movement as devoted to eliminating the arts and other electives from schools. For many, "back to basics" was of a piece with the broader politics of fiscal austerity and moral traditionalism.

If "back to basics" was politicized, the politics were undoubtedly cross-cutting and complicated. For one thing, it is certainly possible to be conservative in matters of education but progressive in other domains. Furthermore, the calls for a return to tradition and basics were not primarily "restorationist," and the promotion of tradition here was not a rejection of modern society but a way of coping with it.[65] That is, those calling for basics did not desire to return to the actual pre-SMSG educational configuration but to remove mathematicians, federal bureaucrats, and national politicians from positions of authority concerning the curriculum. Among the many explanations for the "rise of the right," the experience of the new math points to the importance of shifts in sources of authority.[66] Between the late 1950s and the late 1970s, Americans increasingly looked away from elite officials and academic experts, and toward more lay sources of intellectual and moral authority. Shackelford, the chemist appalled at his daughter's textbook, began his crusade with what would become a characteristic move for critics: contacting the school board and demanding a hearing on the curriculum. "Back to basics" proponents desired the return of power to teachers, parents, and local school boards. The arguments of Shackelford and other critics provide evidence of how the 1970s might be considered the decade of the neighborhood: small groups, local organizations, and nonexperts working to take back control from centralized, elite authorities. This conception did not strictly fall along party lines—local power movements had backers on both the left and the right—and even among conservatives, "back to basics" sat uncomfortably alongside the push for accountability, with supporters divided on whether standards should be set by school boards, states, or the federal government.[67] In this sense, the new math backlash should be situated among uprisings over busing in Boston and NIMBY protests in Orange County, California.[68] These were at once conservative causes *and* manifestations of increasing skepticism toward centralized authority and elite knowledge. Perhaps the only clear party politics angle of "back to basics" was the falling away of any liberal consensus that might have characterized the political culture of the late 1950s and early 1960s.[69]

The new math controversy was nevertheless fundamentally political in the sense that it was about the principles that organize the mind, the family, the society, and the nation. Tellingly, the dean of the College of Education at the University of Illinois explained "back to basics" by noting that "the new math in schools was a symptom of society's disintegration to the public. . . . It was not so much a fear of something new as a fear of losing something old. There was a reluctance to change something comfortable."[70] Alonzo Crim, superintendent of Atlanta public schools in the 1970s, explained the motivation of the new math's critics by noting that "some people are looking for greater regimentation. . . . As they view society in somewhat of a shambles, they feel a more conservative approach is better preparation for their young people."[71] The math curriculum, insofar as it provided a mechanism to ground debates about mental discipline, was ultimately a venue for debates about social order.

Indeed, to critics of the new math, the 1970s seemed to demand different mental habits. The NSF researchers who interviewed math teachers surmised that the continued emphasis on basics was driven by a commitment to the "socializing value" of rote learning. Arithmetic drills were a particularly resonant example of this impulse, as it was widely acknowledged that there was little direct utility in memorization or rapid recall of thousands of arithmetic facts (although it was, of course, always useful to give the right answer). Drilling, however, did foster highly structured, often militaristic classrooms. One teacher explained to the researchers that the study of arithmetic should encourage the development of a "work ethic: responsibility, diligence, persistence, thoroughness, neatness." The chair of an urban high school's math department bluntly rejected any other use for mathematics instruction: "What I tell my classes is this: the only practical value you'll get out of studying mathematics is to learn to do as you're told."[72] The emergence of "back to basics" certainly underscores the remoteness of midcentury fears that learning arithmetic through rote memorization of facts and rules would promote habits of thought too uncomfortably close to those exhibited by "brain-washed" masses deferring to authoritarian pronouncements. Proponents of the new math had endlessly repeated the benefits of possessing the intellectual discipline that enabled citizens to be creative yet analytical in their thought, able to translate the messy physical world into the rigorous language and concepts of mathematics, and then translate answers back into practical action. In turn, criticism of the new math did not just emerge from disillusionment with that premise but also from the argument that desirable intellectual and moral training should involve sources other than professional mathematicians and goals other than thinking structurally.

The issue was not reducible to a simplistic dichotomy between the backers of rote learning and those of conceptual learning. Mathematicians, teachers, and citizens have never agreed completely on the proper balance between those two poles, but few, if any, have thought that it was a matter of only one or the other. That sort of easy distinction is a red herring, drawing attention away from the more important assumption that learning mathematics counts as learning to think. One's view of the new math was understood to be a view about the desirability of elite, technical, structural knowledge in the shaping of American minds.

*　　*　　*

How had the new math become a roundly condemned failure only a decade after its widely hailed—if hastily executed—debut? Parents and teachers still thought of math class as a place where mental habits were cultivated. By the mid-1970s, however, the new math had become emblematic of academic, elite knowledge that had failed to fix the country's ailments. As with many broader rejections of "structural thinking," new math opponents denied the claims of SMSG that learning "modern" mathematics would structure the mind to think more rigorously.[73] The strange notation and curious exercises that had been accepted as part of the new understanding of mathematics were now rejected as parents and teachers called for a return to the traditional practices of memorized facts and rote calculations.

Evaluations of the new math that attribute its decline mainly to unprepared teachers or falling test scores ultimately miss the point. The criticism, from teachers, parents, and concerned citizens alike, was always broader. Behind and beyond evaluations of "modern" math were assumptions about moral and intellectual discipline.

The math classroom remained the place where students learned to think and to be citizens prepared to face the challenges of the modern world. It was not the only source of this training, of course, but for children and young adults it was an influential source. "Textbooks mold nations," the credo of the Gablers, would have been as sensible to proponents of SMSG's curriculum reform in the 1950s as it was to conservatives in the 1970s.[74] Begle and other new math promoters had envisioned students trained in the structure of mathematical knowledge, able to use this sort of reasoning to solve complex and unforeseeable problems. "Back to basics" supporters advocated mathematics training based in discipline and tradition but nonetheless geared to prepare students to meet the challenges of the modern world. In both cases, solutions to the problem of training individual minds were solutions to the problems of social order.

Epilogue

Opponents of the new math won. The curriculum, by most measures, didn't survive the 1970s. Even prior to the backlash, SMSG had closed its own doors in 1972; Edward Begle would be dead of emphysema by 1978. "New math" itself became a slur, a way of labeling dodgy calculations or questionable reforms.

This fate explains why evidence of the new math gleaned from oral histories is exceptionally difficult to use responsibly. Certainly, those who were involved as students, teachers, or authors are often happy to talk about the new math, their willingness perhaps evidence of the importance of the curriculum as a cultural marker of the period. Using such reminiscences as historical evidence, however, would be problematic: the fate of the new math occluded people's memories of its actual origins and deployment. The new math was not a "dead on arrival" reform, too ambitious for schools to deploy successfully; it was not primarily a progressive program of the 1960s; it was not a product of mathematicians run amok or designed without the input of teachers. That it might now be caricatured in these ways suggests the achievement of its opponents.

The curriculum, however, did not just disappear. Texts based on SMSG's material and designed by SMSG-affiliated teachers and mathematicians were widely used until the end of the century. Few publishers returned to the purely formulaic or heavily rule-bound approaches prevalent prior to the reforms. The new math's introduction of sets and structures persisted, although publishers removed "new" or "modern" from textbook titles. Even a 1981 algebra textbook, touted in the preface as "traditional in nature, with very little use of sets and no reference to set properties or operations," nevertheless opened with a discussion of how important sets were to the history of mathematics and began its first chapter with an analysis of the various properties of addition from which the structure of algebra would be built. It remained standard practice in post-new-math algebra textbooks to begin with a discussion of the set of real numbers and the properties of arithmetical operations, and only then to introduce the rules and techniques of algebra. This was very different from the typical pre-new-math algebra textbook's focus on "literal numbers," symbolic manipulation, and formulas, topics that were presented as foundational.[1] The swing "back to basics" was never fully completed—books were altered but only in gradual, piecemeal fashion.

Curriculum reforms are "sticky," after all. Schools can't afford to completely update their materials every time a new textbook is published, and the new math books had a much longer shelf life than public rhetoric about them suggests.[2] Just as importantly, many people involved with groups like SMSG, deeply engaged with the possibilities and opportunities of curriculum reform, never stopped trying to figure out the best way to teach mathematics. This group included successful commercial authors such as Mary. Dolciani, as well as experienced reformers such as James Wilson, Jeremy Kilpatrick, Edwin Moise, Robert Davis, Henry Pollak, and Joseph N. Payne. The last efforts of SMSG, in fact, were largely geared toward measuring the effects of different pedagogical approaches, particularly through longitudinal studies, in some ways laying the groundwork for much subsequent research in math education. The new math, as the first large-scale effort to actually rewrite the nation's textbooks—and measure the results—spawned a generation of people committed to thinking deeply about the role of mathematics in the classroom.

Indeed, the new math was one of the opening volleys in what would become known as the "math wars."[3] Declining test scores continued (and still continue) to be used as evidence of the failure of the latest reform, with blame predictably cycling between an overemphasis on computation skills and an overemphasis on conceptual understanding. By the late 1970s, the falling cost (and size) of electronic calculators and computers again raised questions about the purpose and nature of math class, particularly after the National Council of Teachers of Mathematics (NCTM) recommended that computers and calculators be available to students.[4] The terms of the debate echoed those of the new math: Should math class involve rote exercises and memorization or enable students to discover the structure of knowledge? Is the math curriculum an introduction to how mathematicians think or simply a way to convey the basic facts and techniques of the field? The questions were never about whether students should learn to compute or should learn mathematics in the first place, but about how and why they learned mathematics.

By the 1980s, reformers began pushing back against the "back to basics" movement not by a call for a "new new math" but rather by appropriating the language of "basics" and "standards." NCTM led the charge, arguing that "basics" should include skills other than rote computation and that "standards" should focus on promoting general problem-solving ability. Other groups took up the call of NCTM's 1980 Agenda for Action, particularly the University of Chicago School Mathematics Project.[5]

Most significant, though, was undoubtedly the 1989 release of NCTM's *Curriculum and Evaluation Standards*, which established "values" for "judging the quality" of the mathematics curriculum. The *Standards* followed upon influential articles such as Max Bell's 1974 account of what "everyman should know." While clearly following SMSG's lead, both Bell's article and the *Standards* attempted to shape curricula through establishing benchmarks rather than writing model textbooks.[6] NCTM published similar, though less influential, standards for teaching in 1991 and for assessment in 1995.

The *Standards* both echoed and challenged SMSG's claims about the role of mathematics in American liberal education. The guidelines were meant to provide the goals that students "must master if they are to be self-fulfilled, productive citizens in the next century." The authors posited the challenge of defining "what it means to be mathematically literate . . . in a world where mathematics is rapidly growing and is extensively being applied in diverse fields." Like the new math, the *Standards* drew attention to the need for math class to intellectually prepare citizens for an unknown future.[7] Yet, the *Standards* were not simply an updated new math. Although the document continued to deemphasize the importance of rote memorization, its authors listed SMSG's beloved concept of "mathematics as structure" only as the last of the high school standards and did not even mention the concept as a goal for the lower grades. Furthermore, educators and professors of education (some with extensive math backgrounds), not mathematicians, were in charge of both NCTM and the *Standards* writing committees.

As with SMSG, those promoting the *Standards* had to navigate complicated political pressures. The *Standards* themselves cited the *Nation at Risk* report of 1983, an exemplar of conservative complaints about the "failure" of liberal reforms to improve the schools. NCTM largely self-financed the *Standards* because the group couldn't get national funding in an increasingly conservative political climate whose advocates (at least in rhetoric) demanded that the federal government reduce its educational footprint.[8] Nevertheless, many conservatives initially praised the *Standards* as an example of the sort of "free market" and "voluntary" work by nongovernmental organizations committed to educational improvement. The *Standards* were also well timed to play into President George H. W. Bush's desire for national educational benchmarks, and over forty states subsequently revised their own frameworks. By promoting the need for rigorous "national standards" while simultaneously expanding the notion of "basic" to include far more than rote computational skills, the *Curriculum and Evaluation Standards* remains a record of the educational politics of the 1980s.[9]

Despite such attempts at balance and initial praise by Republican elites, the *Standards* quickly faced critics who claimed that reformers were once again misguidedly moving the emphasis away from computational proficiency. Labeling NCTM's later reforms "new new math," critics also accused NCTM of focusing too much on a "choose your own adventure" style of mathematics progressivism; one opposition group's motto was simply "there is a mathematically correct solution."[10] Certainly the *Standards* were not perfect documents. From debates within the writing teams to later skirmishes over revisions, setting out national goals for mathematics class is a fraught exercise.[11] It is worth underscoring, however, how much the often histrionic debate further distorted the memory of the new math: to imply that opposing the "new math" meant promoting "mathematically correct solutions" was to fundamentally misrepresent the history of curricular reform.

*   *   *

The mastery of a technical subject always requires both conceptual understanding and routine practice. Yet the relative importance ascribed to these components implies judgments about the purpose and nature of learning the subject. It shouldn't be surprising that the issues surrounding the NCTM *Standards*—and in the twenty-first century, those surrounding the Common Core State Standards—harkened back to those surrounding the new math. If learning math counted as learning to think, then claims about how one should learn math entailed decisions about the desirability of particular mental habits.

The point of the new math may have been to shape Americans' habits of thought, but the chosen mechanism was a specific brand of "modern" mathematics. SMSG itself was a state-sponsored program, with the imprimatur of professional mathematicians, to develop a "modern" mathematics curriculum in the service of addressing apparent problems of mental discipline. The math classroom was believed to instantiate and inculcate intellectual rigor, and the curriculum's evaluation depended on changing opinions about who should be trusted to specify how students learn to reason.

The politicization of schools and classrooms provides an opportunity for historians. Schools train students about what is out there in the world and train them to navigate that world. Otherwise-abstract intellectual conceptions are made concrete in—and their implications understood through—the curriculum. Historians, in turn, can learn more about how Americans understood, resisted, or embraced notions of discipline and rigor, reason and rationality, liberalism and conservatism, by tracing how they responded to the curriculum. Past controversies surrounding the curriculum reveal not just the way millions of people made sense of claims about "right thinking" or desirable intellectual habits, but also how they connected those claims to larger political movements.

The history of the new math should also prompt those who write, use, or evaluate textbooks to remember that the design of the math curriculum is never just a matter of deciding which topics to cover or which pedagogical techniques raise test scores. The constellation of factors that made the new math unique—direct involvement of federal authorities and monies in textbook development, leading roles played by academic mathematicians, Cold War concerns about authoritarian personalities, midcentury claims about the underlying nature of mathematical knowledge—may be period specific. Yet the math classroom will remain a political venue as long as learning math counts as learning to think. Debates about the American math curriculum are debates about the nature of the American subject.

Acknowledgments

It is a pleasure to acknowledge the assistance I've received over the years, even if I can't possibly mention everyone who supported this endeavor. First and foremost, Steven Shapin has been a constant source of wisdom—and a model scholar—for me. He has encouraged and helped shape this project from the very beginning and I am greatly indebted to him. Charles Rosenberg and Lizabeth Cohen have patiently and carefully waded through multiple revisions, adding perceptive and valuable feedback at each stage. A few other readers were also subjected to the entire manuscript, and to them I owe a very special thanks: Jamie Cohen-Cole, Stephanie Dick, Moon Duchin, David Kaiser, Ashley Newman-Owens, Mary Ann Phillips, David Roberts, Alma Steingart, and Nasser Zakariya. Many colleagues over the years, particularly Jeremy Blatter, Paul Cruickshank, Meg Formato, Andrew Jewitt, Heidi Voskuhl, and Alex Wellerstein, helped shape and refine one aspect or another of the book. At the University of Chicago Press, Christie Henry, Abby Collier, Amy Krynak, Mary Corrado, and Kathryn Gohl ably shepherded the project through its final stages. John Rudolph and Ted Porter provided careful input on the whole manuscript, undoubtedly strengthening the final product. Any remaining faults are, of course, entirely mine.

For institutional support, I thank the faculty and staff both of the NYU Gallatin School of Individualized Study, where the book was completed, and of Harvard's History of Science Department, where much of the book was conceived and written. Financial support was provided by the Charles Warren Center for Studies in American History, Harvard University's Center for American Political Studies, the Erwin Hiebert Fund for Graduate Research and Travel, and an Ashford Family Dissertation Completion Fellowship. Research assistance has been provided by the staff of Widener Library, Harvard University; Monroe C. Gutman Library, Harvard Graduate School of Education; Dolph Briscoe Center for American History, University of Texas at Austin; Sullivant Library, Ohio State University; and the school committees of Newton, Belmont, and Boston, Massachusetts. In this capacity, I especially thank Fred Burchsted (Widener Library) and Edward Copenhagen (Gutman Library).

Some parts of the book have appeared in slightly different form in the *Journal of American History* and *Isis*. For comments on these sections in particular, I also thank Andrew Hartman, John M. Heffron, Bernard Lightman, Edward Linenthal, and the anonymous referees.

Finally, I want to acknowledge my family. Mary Ann Phillips and Shane Phillips have always been enthusiastic about my work, however far removed from their own lives, and I am so deeply grateful for them. Anna Evans helped me to finish this book in more ways than she will ever fully realize, and I am truly appreciative of her presence in my life. To them, and to the memory of Jim Phillips and Erin Aylward, this book is dedicated.

Notes

Chapter 1. Introduction

1. Harry Schwartz, "The New Math Is Replacing Third 'R,'" *New York Times*, January 25, 1965, 18.

2. Speech, in folder "Speeches—E. G. Begle," box 13.4/86–28/48, School Mathematics Study Group Records, 1958–1977 (hereafter SMSGR), Archives of American Mathematics, Dolph Briscoe Center for American History, University of Texas at Austin.

3. SMSG, "Gross Sales Report," in folder "Advisory Committee—Replies," box 4.1/86–28/1, SMSGR.

4. E. G. Begle, letters of July 25 and 31, 1958, in folder "National Science Foundation General," box 13.5/86–28/59, SMSGR.

5. E. G. Begle, "The School Mathematics Study Group," *NASSP Bulletin* 43 (May 1959): 26–31, at 27–28.

6. Some historical work has been done on the general institutional history of new math organizations: Robert W. Hayden, "A History of the 'New Math' Movement in the United States" (Ph.D. diss., Iowa State University, 1981); David L. Roberts and Angela Walmsley, "The Original New Math: Storytelling versus History," *Mathematics Teacher* 96 (2003): 468–73; Angela L. E. Walmsley, "A History of the 'New Mathematics' Movement and Its Relationship with the National Council of Teachers of Mathematics Standards" (Ph.D. diss., Saint Louis University, 2001); George M. A. Stanic and Jeremy Kilpatrick, eds., *A History of School Mathematics*, 2 vols. (Reston, VA: National Council of Teachers of Mathematics, 2003), vol. 1, chaps. 11–17. For an internal history, see William Wooton, *SMSG: The Making of a Curriculum* (New Haven: Yale University Press, 1965).

7. Ian Muller, "Mathematics and Education: Some Notes on the Platonic Program," *Apeiron* 24 (1991): 85–104, esp. 85.

8. Reviel Netz, *The Shaping of Deduction in Greek Mathematics: A Study in Cognitive History* (Cambridge: Cambridge University Press, 1999), 309–12.

9. Plato, *Meno*, in Edith Hamilton and Huntington Cairns, eds., W. K. C. Guthrie, trans., *The Collected Dialogues of Plato including the Letters*, 353–84 (Princeton: Princeton University Press, 1961), at 369–71.

10. Matthew L. Jones, *The Good Life in the Scientific Revolution: Descartes, Pascal, Leibniz, and the Cultivation of Virtue* (Chicago: University of Chicago Press, 2006); cf. Roger Ascham, *The Scholemaster* (London, 1579). For contemporary debates about the relevance of mathematics for the emerging practices of natural philosophy, see Steven Shapin, *A Social History of Truth: Civility and Science in Seventeenth-Century England* (Chicago: University of Chicago Press, 1994), esp. 315–52; and Peter Dear, *Discipline and Experience: The Mathematical Way in the Scientific Revolution* (Chicago: University of Chicago Press, 1995).

11. Montucla, as quoted in Joan L. Richards, "Historical Mathematics in the French Eighteenth Century," *Isis* 97 (2006): 700–713, at 707. Galileo's *The Assayer* contains the most well-known pronouncement that the "book of nature" was written in the language of mathematics: Stillman Drake, ed. and trans., *Discoveries and Opinions of Galileo* (New York: Anchor Books, 1957), 238. For the close relationship of mathematics and reason in the Enlightenment, see also Lorraine Daston, "Enlightenment Calculations," *Critical Inquiry* 21 (1994): 182–202; and L. Daston, *Classical Probability in the Enlightenment* (Princeton: Princeton University Press, 1988). The notion of the history of mathematics as a mirror of the history of human civilization remained prevalent into the twentieth century. See, e.g., Lancelot Hogben, *Mathematics for the Million: A Popular Self Educator* (London: George Allen and Unwin, 1936), 34.

12. E.g., Andrew Warwick, *Masters of Theory: Cambridge and the Rise of Mathematical Physics* (Chicago: University of Chicago Press, 2003).

13. George Elder Davie, *The Democratic Intellect: Scotland and Her Universities in the Nineteenth Century*, 2nd ed. (Edinburgh: Edinburgh University Press, 1964 [1961]), 127; also Richard Olson, "Scottish Philosophy and Mathematics, 1750–1830," *Journal of the History of Ideas* 32 (1971): 29–44.

14. For exemplary historical studies of the American experience of meritocratic rankings, see John Carson, *The Measure of Merit: Talents, Intelligence, and Inequality in the French and American Republics, 1750–1940* (Princeton: Princeton University Press, 2007); and Nicholas Lemann, *The Big Test: The Secret History of the American Meritocracy* (New York: Farrar, Straus and Giroux, 1999).

15. Jo Boaler, "The Stereotypes about Math That Hold Americans Back," *Atlantic*, November 12, 2013, at http://www.theatlantic.com/education/archive/2013/11/ . This is hardly limited to the United States, of course. See, e.g., a 2008 British think tank report titled "The Value of Mathematics" that claimed the discipline "develops the fundamental skills of logical and critical reasoning, training the mind to be highly analytical and to deal with complex problems. It provides the basic language, structures and theories for understanding the world around us": Laura Kounine, John Marks, and Elizabeth Truss, "The Value of Mathemat-

ics," *Reform Report* (February 2008): 6, available at http://www.reform.co.uk/ client_files/www.reform.co.uk/files/the_value_of_mathematics.pdf.

16. Basil B. Bernstein, *Class, Codes, and Control*, 3 vols. (London: Routledge and Kegan Paul, 1971–75); Peter Woods and Martyn Hammersley, eds., *School Experience: Explorations in the Sociology of Education* (New York: St. Martin's Press, 1977), 14; Ian Hunter, "Assembling the School," in Andrew Barry, Thomas Osborne, and Nikolas Rose, eds., *Foucault and Political Reason: Liberalism, Neo-Liberalism, and the Rationalities of Government*, 143–66 (Chicago: University of Chicago Press, 1996), esp. 149–55.

17. Michel Foucault, "About the Beginning of the Hermeneutics of the Self: Two Lectures at Dartmouth" [1980], *Political Theory* 21 (1993): 198–227, at 203.

18. Erwin Panofsky, *Gothic Architecture and Scholasticism*, Wimmer Memorial Lecture, 1948 (Latrobe, PA: St. Vincent Archabbey, 1951). More generally, see Bruce Holsinger, *The Premodern Condition: Medievalism and the Making of Theory* (Chicago: University of Chicago Press, 2005), esp. chap. 3, and Laurence Petit's translation of Bourdieu's "Postface" to Panofsky's lecture, 221–42; also Pierre Bourdieu, "Systems of Education and Systems of Thought," in Michael F. D. Young, ed., *Knowledge and Control: New Directions for the Sociology of Education*, 189–207 (London: Collier Macmillan, 1971); and Pierre Bourdieu and Jean-Claude Passeron, *Reproduction in Education, Society, and Culture*, trans. Richard Nice (London: Sage, 1977).

19. Andrew Warwick and David Kaiser, "Conclusion: Kuhn, Foucault, and the Power of Pedagogy," in David Kaiser, ed., *Pedagogy and the Practice of Science: Historical and Contemporary Perspectives*, 393–409 (Cambridge: MIT Press, 2005), at 401, as well as Warwick, *Masters of Theory*, and David Kaiser, *Drawing Theories Apart: The Dispersion of Feynman Diagrams in Postwar Physics* (Chicago: University of Chicago Press, 2005). Among other examples of the increasing role of pedagogy in historical accounts of physics in particular, see Suman Seth, *Crafting the Quantum: Arnold Sommerfield and the Practice of Theory, 1890–1926* (Cambridge: MIT Press, 2010); and Kathryn M. Olesko, *Physics as a Calling: Discipline and Practice in the Königsberg Seminar for Physics* (Ithaca, NY: Cornell University Press, 1991).

20. James Bryant Conant, *Education and Liberty: The Role of the Schools in a Modern Democracy* (Cambridge: Harvard University Press, 1953), 2. More generally, Michael B. Katz, *The Irony of Early School Reform: Educational Innovation in Mid-Nineteenth Century Massachusetts* (Cambridge: Harvard University Press, 1968); Stephen Thernstrom, *Poverty and Progress: Social Mobility in a Nineteenth-Century City* (Cambridge: Harvard University Press, 1981); Claudia Goldin and Lawrence Katz, "Human Capital and Social Capital: The Rise of Secondary Schooling in America, 1910–1940," *Journal of Interdisciplinary History* 29 (1999): 683–723.

21. On other possible theoretical approaches and orientations to citizenship,

see Derek Heater, *A Brief History of Citizenship* (New York: New York University Press, 2004); and Mae M. Ngai, *Impossible Subjects: Illegal Aliens and the Making of Modern America* (Princeton: Princeton University Press, 2004).

22. On "Cold War rationality," see Paul Erickson et al., *How Reason Almost Lost Its Mind: The Strange Career of Cold War Rationality* (Chicago: University of Chicago Press, 2013); on the relationship between government and the human sciences, Nikolas Rose, *Governing the Soul: The Shaping of the Private Self*, 2nd ed. (London: Free Association Books, 1999); and historically, Jan Goldstein, *The Post-Revolutionary Self: Politics and Psyche in France, 1750–1850* (Cambridge: Harvard University Press, 2005), and Charles Taylor, *Sources of the Self: The Making of the Modern Identity* (Cambridge: Harvard University Press, 1989). The "American self" at midcentury is briefly discussed below, in chap. 4, and also examined in Jamie Cohen-Cole, *The Open Mind: Cold War Politics and the Sciences of Human Nature* (Chicago: University of Chicago, 2014); Ron Robin, *The Making of the Cold War Enemy* (Princeton: Princeton University Press, 2001); and Ellen Herman, *The Romance of American Psychology: Political Culture in the Age of Experts* (Berkeley: University of California Press, 1995), esp. chap. 4.

23. E.g., Bob Moon, *The "New Maths" Curriculum Controversy* (London: Falmer Press, 1986); Organization for European Economic Co-Operation, Office for Scientific and Technical Personnel, *New Thinking in School Mathematics* (May 1961); David Rappaport, "The Nuffield Mathematics Project," *Elementary School Journal* 71 (1971): 295–308; Société mathématique de France, *Structures algébraiques et structures topologiques* (Geneva: Enseignement mathématique, 1958). For SMSG's translation project, see "Spanish Translations," box 86–28/47, SMSGR.

24. Carl F. Kaestle, *Pillars of the Republic: Common Schools and American Society, 1780–1860* (New York: Hill and Wang, 1983), 97. The literature on this point is vast. Some classic texts on the "social role" of American schools include the following: Lawrence A. Cremin, *The American Common School: An Historic Conception* (New York: Teachers College, 1951); Bernard Bailyn, *Education in the Formation of American Society* (Chapel Hill: University of North Carolina Press, 1960); David Tyack and Elisabeth Hansot, *Managers of Virtue: Public School Leadership in America, 1820–1980* (New York: Basic Books, 1982); Joel Spring, *The American School: 1642–1993*, 3rd ed. (New York: McGraw-Hill, 1994); and William J. Reese, *America's Public Schools: From the Common School to "No Child Left Behind"* (Baltimore: Johns Hopkins University Press, 2005). For the university: Laurence R. Veysey, *The Emergence of the American University* (Chicago: University of Chicago Press, 1965); and Julie A. Reuben, *The Making of the Modern University: Intellectual Transformation and the Marginalization of Morality* (Chicago: University of Chicago Press, 1996).

25. "Third Draft of the SMSG Position Statement" [June 29, 1962], box 86–28/45, SMSGR, 2.

26. The best history of midcentury curriculum reform in the sciences is John

Rudolph, *Scientists in the Classroom: The Cold War Reconstruction of American Science Education* (New York: Palgrave, 2002).

27. Edward G. Begle, "Review of *Why Johnny Can't Add*," *National Elementary Principal* 53 (January–February 1974): 26–31, at 27.

28. Begle, as quoted in Frank Kendig, "Does the New Math Add Up?" *New York Times*, January 6, 1974, E14, 28, 32.

29. An early draft of the published report was used by SMSG working groups and was later published as College Entrance Examination Board, *Report of the Commission on Mathematics: Program for College Preparatory Mathematics* (New York: College Entrance Education Board, 1959).

30. Thomas Steven Dupre, "The University of Illinois Committee on School Mathematics and the 'New Mathematics' Controversy" (Ph.D. diss., University of Illinois, 1986), esp. 24–29; B. E. Meserve, "The University of Illinois List of Mathematical Competencies," *School Review* 61 (1953): 85–92; and M. Beberman, *An Emerging Program of Secondary School Mathematics* (Cambridge: Harvard University Press, 1958).

31. For a contemporary listing of some of the major mathematics curriculum projects in operation, see National Council of Teachers of Mathematics, *An Analysis of New Mathematics Programs* (Washington, DC: NCTM, 1963), as well as H. Victor Crespy, "A Study of Curriculum Development in School Mathematics by National Groups, 1950–1966: Selected Programs" (Ph.D. diss., Temple University, 1969).

32. Cambridge Conference on School Mathematics, *Goals for School Mathematics* (Boston: Houghton Mifflin, 1963). Other reformers were similarly pushing for more substantial changes to the nature of math class, e.g., Robert B. Davis, *The Changing Curriculum: Mathematics* (Washington, DC: Association for Supervision and Curriculum Development, NEA, 1967).

33. E. G. Begle, "Some Lessons Learned by S.M.S.G.," *Mathematics Teacher* 66 (1973): 207–14; Wooton, *SMSG*, 42; interview with Paul Rosenbloom, transcript, box 4RM19, R. L. Moore Legacy Collection (hereafter Moore Collection), 1890–1900, 1920–2009, Archives of American Mathematics, Dolph Briscoe Center for American History, University of Texas at Austin, pp. 44–45. This interview forms part of a set conducted by historians around the turn of the twenty-first century, dedicated to exploring the way mathematicians conceive pedagogical reform generally. Because the collection involves mathematicians' recollections many years after the fact, I use it sparingly, but the interviews provide some details not otherwise available.

34. Tom Loveless, ed., *The Great Curriculum Debate: How Should We Teach Reading and Math?* (Washington, DC: Brookings Institution Press, 2001), as well as Jeanne S. Chall, *Learning to Read: The Great Debate*, 3rd ed. (Fort Worth: Harcourt Brace College Publishers, 1996). On contemporary assessments of "back to basics" in English, see *English Journal* 65 (November 1976).

35. Morris Kline, *Why Johnny Can't Add: The Failure of the New Math* (New York: Vintage Books, 1973); Rudolph Flesch, *Why Johnny Can't Read—And What You Can Do about It* (New York: Harper and Brothers, 1955).

Chapter 2. The Subject and the State

1. The earliest reference to "Cold War of the classrooms" in this sense was likely the testimony of Lewis Strauss, of the Atomic Energy Commission. See "Cold War of the Classrooms," *Science Digest* 39 (February 1956): 33.

2. Lawrence A. Cremin, *The Genius of American Education* (New York: Vintage Books, 1965), 11.

3. E.g., Robert A. Divine, *The Sputnik Challenge* (New York: Oxford University Press, 1993), esp. 5–6, 18–37; Paul Dickson, *Sputnik: The Shock of the Century* (New York: Berkley Books, 2001), esp. 10–11, 84–85. Many other elements of the launch created fear as well, from the first violation of continental air space in the nation's history to the satellite's mysterious ability to open garage doors: see Dickson, *Sputnik*, 96, 110–18, 128. The most direct consequence of the launch was, not surprisingly, the rapid development of U.S. missile and space programs: Kendrick Oliver, *To Touch the Face of God: The Sacred, the Profane, and the American Space Program, 1957–1975* (Baltimore: Johns Hopkins University Press, 2012); Walter A. McDougall, *The Heavens and the Earth: A Political History of the Space Age* (New York: Basic Books, 1985).

4. An indispensable resource for these in some ways parallel reforms is John Rudolph, *Scientists in the Classroom: The Cold War Reconstruction of American Science Education* (New York: Palgrave, 2002).

5. Doyle Bortner, "A Study of Published Lay Opinion on Educational Programs and Problems," *Education* (June 1951): 644, as cited in Adam Benjamin Golub, "Into the Blackboard Jungle: Educational Debate and Cultural Change in 1950s America" (Ph.D. diss., University of Texas at Austin, 2004), 20; Golub reads this study as already outdated, given that the media coverage (but perhaps not popular opinion) of the schools turned increasingly negative during the early 1950s; survey data from National Opinion Research Center, University of Chicago, March 12– 22, 1955. All survey data retrieved from the iPOLL Databank, Roper Center for Public Opinion Research, University of Connecticut, http://www.ropercenter.uconn.edu; Arthur Zilversmit, *Changing Schools: Progressive Education Theory and Practice, 1930–1960* (Chicago: University of Chicago Press, 1993), 114; Rudolph, *Scientists in the Classroom*, 17; David Tyack and Larry Cuban, *Tinkering toward Utopia: A Century of Public School Reform* (Cambridge: Harvard University Press, 1995), 32. For an example of the emphasis on gradual reform, Roy Larsen and James B. Conant joined to form the National Citizens Commission for Public Schools after the war, an organization de-

voted to working with school teachers, parents, and administrators to reform schools from within: Roy E. Larsen Collection, Special Collections, Monroe C. Gutman Library, Harvard Graduate School of Education.

6. James B. Conant, *The American High School Today: A First Report to Interested Citizens* (New York: McGraw-Hill, 1959), esp. 37–40; William J. Reese, *America's Public Schools: From the Common Schools to "No Child Left Behind"* (Baltimore: Johns Hopkins University Press, 2005), 253–54.

7. Gilbert E. Smith, *The Limits of Reform: Politics and Federal Aid to Education, 1937–1950*, updated ed. (New York: Longman, 1989).

8. Bailey is given credit by James W. Guthrie, "A Political Case History: Passage of the ESEA," *Phi Delta Kappan* 49 (1968): 302–6, at 302; Barbara Barksdale Clowse, *Brainpower for the Cold War: The Sputnik Crisis and National Defense Education Act of 1958* (Westport, CT: Greenwood Press, 1981), 61–92.

9. Reese, *America's Public Schools*, 215. Schools themselves were conservative institutions, but education reform rhetoric cycles rapidly as the politics of education change. See Tyack and Cuban, *Tinkering toward Utopia*, 42; Larry Cuban, *How Teachers Taught: Constancy and Change in American Classrooms, 1880–1990*, 2nd ed. (New York: Teachers College Press, 1993), esp. 239.

10. For *Brown v. Board of Education*, see James T. Patterson, *Brown v. Board of Education: A Civil Rights Milestone and Its Troubled Legacy* (New York: Oxford University Press, 2001); Mac A. Stewart, ed., *The Promise of Justice: Essays on Brown v. Board of Education* (Columbus: Ohio State University Press, 2008); and Michael J. Klarman, *Brown v. Board of Education and the Civil Rights Movement* (New York: Oxford University Press, 2007).

11. Ronald Lora, "Education: Schools as Crucible in Cold War America," in Robert H. Bremner and Gary W. Reichard, eds., *Reshaping America: Society and Institutions, 1945–1960*, 223–60 (Columbus: Ohio State University Press, 1982); Jonathan Zimmerman, *Whose America? Culture Wars in the Public Schools* (Cambridge: Harvard University Press, 2002); Jerome Karabel, *The Chosen: The Hidden History of Admission and Exclusion at Harvard, Yale, and Princeton* (New York: Houghton Mifflin, 2005); Nicholas Lemann, *The Big Test: The Secret History of the American Meritocracy* (New York: Farrar, Straus and Giroux, 1999); John Gardner, *Excellence: Can We Be Equal and Excellent Too?* (New York: Harper and Row, 1961); and more generally on the changing nature of public institutions at midcentury, Lizabeth Cohen, *A Consumers' Republic: The Politics of Consumption in Postwar America* (New York: Vintage Books, 2004), esp. 257–89, and Kenneth Jackson, *Crabgrass Frontier: The Suburbanization of the United States* (New York: Oxford University Press, 1985); see also chap. 6, below.

12. The voluminous hearings were published as Congress, Senate, Committee on Armed Services, *Inquiry into Satellite and Missile Programs Part 1*, 85th Cong., 1st and 2nd sess., November 25–27; December 13, 14, 16, 17; January 10,

13, 15–17, 20, 21, 23; *Part 2*, 85th Cong., 2nd sess., January 6, 8–10, 13–17, 20–22, 1958; *Part 3*, February 26, April 3, and July 24, 1958. Bush: Senate, *Inquiry into Satellite and Missile Programs*, 64; Doolittle: *Inquiry into Satellite and Missile Programs*, 112, 124.

13. H. G. Rickover, *Education and Freedom* (New York: Dutton, 1959), 35 and 38 (based on lectures given during 1956–58).

14. Fulbright: Congress, Senate, 85th Cong., 2nd sess., *Congressional Record* 104, pt. 1 (January 23, 1958): 872; Zelenko: Congress, House, 85th Cong., 2nd sess., *Congressional Record* 104, pt. 13 (August 8, 1958): 16685; Hill: Congress, Senate, 85th Cong., 2nd sess., *Congressional Record* 104, pt. 13 (August 13, 1958): 17235, and Senate, Committee on Labor and Public Welfare, *Science and Education for National Defense*, 85th Cong., 2nd sess., January 21, 23, 28–30; February 6, 7, 18–21, 24–27; March 3, 5, 6, 10–13, 1958, at 2.

15. Hoover: Henry Toy, Jr., radio broadcast no. 10, December 8, 1957, transcript in vol. IVb of National Citizens Council for Better Schools, *Highlights Report* 1957, box 4, Roy E. Larsen Collection, Special Collections, Monroe C. Gutman Library, Harvard Graduate School of Education; Flanders: Senate, *Inquiry into Satellite and Missile Programs*, 55; Phillips: Senate, *Inquiry into Satellite and Missile Programs*, 616; Mansfield: Senate, *Science and Education*, 881; Gwinn: Congress, House, 85th Cong., 2nd sess., *Congressional Record* 104, pt. 1 (January 15, 1958): 496–97.

16. Flanders's speech given February 21, 1958, and later published in Congress, Senate, "Education for Survival," 85th Cong., 2nd sess., *Congressional Record* 104, pt. 2 (February 25, 1958): 2683–85.

17. An excellent source on the rise and fall of progressive education still remains Lawrence A. Cremin, *The Transformation of the School: Progressivism in American Education, 1876–1957* (New York: Alfred A. Knopf, 1969); for the fate of educational progressives in the 1950s, see Andrew Hartman, *Education and the Cold War: The Battle for the American School* (New York: Palgrave Macmillan, 2008); Rudolph, *Scientists in the Classroom*, chap. 1. Dewey argued that pedagogy must utilize the social surroundings of the child to shape his or her education. His actual pedagogical writings are voluminous but opaque. See John Dewey, "My Educational Creed," 3–18, in J. Dewey, *Teachers Manuals* (New York: E. L. Kellogg, 1899); J. Dewey, *The School and Society*, 2nd rev. ed. (Chicago: University of Chicago Press, 1947) (lectures originally given in 1899); J. Dewey, *The Child and the Curriculum* (Chicago: University of Chicago Press, 1902); as well as Robert B. Westbrook, *John Dewey and American Democracy* (Ithaca, NY: Cornell University Press, 1991). Dewey's critics and some of his supporters unhelpfully have lumped his views together with those of Johann Friedrich Herbart and Jean-Jacques Rousseau, among others. For a more recent example of the ongoing progressive education debate, see E. D. Hirsch Jr., "The Roots of the Education Wars," in Tom Loveless, ed., *The Great Curric-*

ulum Debate: How Should We Teach Reading and Math? 13–24 (Washington, DC: Brookings Institution Press, 2001).

18. Ellen Condliffe Lagemann, *An Elusive Science: The Troubling History of Education Research* (Chicago: University of Chicago Press, 2000), esp. pt. 2; Cremin, *Transformation of the School*, 274–75.

19. Flanders: Senate, "Education for Survival," 2683.

20. James Bryant Conant, *The Child, the Parent, and the State* (Cambridge: Harvard University Press, 1959), chap. 4. Claudia Goldin and Lawrence Katz have more recently looked at the intersection of economics and education in this period and beyond; see, for example, Claudia Goldin and Lawrence F. Katz, *The Race between Education and Technology* (Cambridge: Belknap Press of Harvard University Press, 2008).

21. Cremin, *Transformation of the School*, 324–28, 347–49.

22. Robert Hutchins, *The Conflict in Education in a Democratic Society* (New York: Harper and Brothers, 1953), 22, 69–70, and 95. Hutchins was one of the prime forces behind the promotion of "great books" curricula and self-instruction programs: Alex Beam, *A Great Idea at the Time: The Rise, Fall, and Curious Afterlife of the Great Books* (New York: Public Affairs, 2008).

23. Arthur Bestor, *Educational Wastelands: The Retreat from Learning in our Public Schools*, 2nd ed. (Urbana: University of Illinois Press, 1985 [1953]); Rudolph, *Scientists in the Classroom*, chap. 1. For the range of criticism, see Howard Ozmon, ed., *Contemporary Critics of Education* (Danville, IL: Interstate Printers and Publishers, 1970).

24. Bestor, *Educational Wastelands*, 18–21; Lagemann, *An Elusive Science*, 160; Clowse, *Brainpower for the Cold War*, 23.

25. Hutchins, *Conflict in Education*, 20–21.

26. Bestor, *Educational Wastelands*, 179–87.

27. Ellen Schrecker, *No Ivory Tower: McCarthyism and the Universities* (New York: Oxford University Press, 1986); Diane Ravitch, *The Troubled Crusade: American Education, 1945–1980* (New York: Basic Books, 1983), 81–113; Golub, "Into the Blackboard Jungle," chap. 2; Richard H. Pells, *The Liberal Mind in a Conservative Age: American Intellectuals in the 1940s and 1950s* (New York: Harper and Row, 1985), esp. chap. 5. On the range of responses to communism, see Michael Kimmage, *The Conservative Turn: Lionel Trilling, Whittaker Chambers, and the Lessons of Anti-Communism* (Cambridge: Harvard University Press, 2009); for scientists and anticommunism, Jessica Wang, *American Science in an Age of Anxiety: Scientists, Anticommunism, and the Cold War* (Chapel Hill: University of North Carolina Press, 1999), and the many studies of the Oppenheimer security hearings, e.g., Charles Thorpe, *Oppenheimer: The Tragic Intellect* (Chicago: University of Chicago Press, 2006), chap. 7, and Kai Bird and Martin J. Sherwin, *American Prometheus: The Triumph and Tragedy of J. Robert Oppenheimer* (New York: Vintage, 2005), chaps. 31–36.

28. David A. Hollinger, "Science as Weapon in *Kulturekämpfe* in the United States during and after World War II," in D. Hollinger, *Science, Jews, and Secular Culture: Studies in Mid-Twentieth-Century American Intellectual History*, 155–74 (Princeton: Princeton University Press, 1996), 160; more generally, Mark Solovey and Hamilton Cravens, eds., *Cold War Social Science: Knowledge Production, Liberal Democracy, and Human Nature* (New York: Palgrave Macmillan, 2012), esp. pts. 2 and 3; Jamie Cohen-Cole, "The Reflexivity of Cognitive Science: The Scientist as Model of Human Nature," *History of the Human Sciences* 18 (2005): 107–39; J. Cohen-Cole, *The Open Mind: Cold War Culture and the Sciences of Human Nature* (Chicago: University of Chicago Press, 2014), chaps. 3–5; David Paul Haney, *The Americanization of Social Science: Intellectuals and Public Responsibility in the Postwar United States* (Philadelphia: Temple University Press, 2008), esp. 95–117.

29. Richard Hofstadter, *Anti-Intellectualism in American Life* (New York: Vintage Books, 1963), esp. pt. 5; Hartman, *Education and the Cold War*, chap. 5, esp. 132–33; Sterling M. McMurrin, "The Curriculum and the Purposes of Education," in Robert Heath, ed., *New Curricula*, 262–84 (New York: Harper and Row, 1964), at 267.

30. Some scientists explicitly justified their efforts as a fight against anti-intellectualism, e.g., Jack S. Goldstein, *A Different Sort of Time: The Life of Jerrold R. Zacharias Scientist, Engineer, Educator* (Cambridge: MIT Press, 1992), 165. There is certainly evidence that teachers resented that the reform funding emerged only after scientists were put in charge. See the letter of Bruce Meserve to Begle, October 12, 1965, in folder "NCTM 1965–1966," box 13.6/86–28/60, School Mathematics Study Group Records, 1958–77 (hereafter SMSGR), Archives of American Mathematics, Dolph Briscoe Center for American History, University of Texas at Austin; as well as an NCTM invitation for criticism of SMSG and the somewhat hostile responses in reply, in folder "NCTM," box 13.5/86–28/59, and folder "NCTM Issues," box 13.6/86–28/76, SMSGR.

31. Rockefeller Brothers Fund, "The Pursuit of Excellence: Education and the Future of America," Special Studies Project Report V, American at Mid-Century series (Garden City, NJ: Doubleday, 1958), v, 49.

32. Committee on the Objectives of a General Education in a Free Society, *General Education in a Free Society: Report of the Harvard Committee* (Cambridge: Harvard University Press, 1945), 54, 75–76. The Red Book was widely publicized: e.g., "U.S. High School," *Life* (April 22, 1946): 87–93, and "New Harvard Plan Backed by Faculty," *New York Times*, November 1, 1945, 25.

33. Education Policies Commission of the National Education Association, American Association of School Administrators, and the Executive Committee of the American Council on Education, *Education and National Security* (Washington, DC: Education Policies Commission and the American Council on Education, 1951), 15, 18.

34. Zacharias: G. K. Hodenfield, "What's New in the Curriculum," *PTA Magazine* 61 (January 1967): 7–9, at 9. More generally, William Whyte, *The Organization Man* (New York: Simon and Schuster, 1956); David Riesman with Reuel Denney and Nathan Glazer, *The Lonely Crowd: A Study of the Changing American Character* (New Haven: Yale University Press, 1950); Vance Packard, *The Status Seekers* (New York: Pocket Books, 1961 [1959]); and for an analysis of these claims in the context of the scientific community, see David Kaiser, "The Postwar Suburbanization of American Physics," *American Quarterly* 56 (2004): 851–88, and Steven Shapin, *The Scientific Life: A Moral History of a Late Modern Vocation* (Chicago: University of Chicago Press, 2008), esp. 119–21.

35. Sloan Wilson, *The Man in a Gray Flannel Suit* (New York: Simon and Schuster, 1955); Ayn Rand, *Atlas Shrugged* (New York: Random House, 1957); more generally, Thomas Doherty, *Cold War, Cool Medium: Television, Mc-Carthyism, and American Culture* (New York: Columbia University Press, 2003); Stephen J. Whitfield, *The Culture of the Cold War*, 2nd ed. (Baltimore: Johns Hopkins University Press, 1996).

36. Philip Wylie, *Generation of Vipers*, new ed. (New York: Pocket Books, 1955); Edward A. Strecker, *Their Mothers' Sons: The Psychiatrist Examines an American Problem* (New York: J. B. Lippincott, 1946); Rebecca Jo Plant, *Mom: The Transformation of Motherhood in Modern America* (Chicago: University of Chicago Press, 2010), esp. 20 and chap. 3.

37. National Congress of Parents and Teachers, *Proceedings: Annual Convention Official Reports and Records* 62 (May 18–21, 1958): 15–16; Gardner, *Excellence*, 145–46.

38. Bosley Crowther, "The Screen: 'Blackboard Jungle'" *New York Times*, March 21, 1955, 21; Golub, "Into the Blackboard Jungle," chap. 3; more generally, James Gilbert, *A Cycle of Outrage: America's Reaction to the Juvenile Delinquent in the 1950s* (New York: Oxford University Press, 1986), and Margaret Peacock, *Cold War Kids: Images of Soviet-American Childhood and the Collapse of Consensus, 1945–1968* (Chapel Hill: University of North Carolina Press, forthcoming), esp. chap. 3.

39. Walter B. Kolesnik, "Mental Discipline and the Modern Curriculum," *Peabody Journal of Education* 35 (March 1958): 298–303, at 302; Thurmond: Congress, Senate, 85th Cong., 2nd sess., *Congressional Record* 104, pt. 13 (August 13, 1958): 17296.

40. The series had no author listed but was published as "Crisis in Education: Schoolboys Point up a U.S. Weakness," *Life* 44 (March 24, 1958): 25–35; "An Underdog Profession Imperils the Schools," *Life* 44 (March 31, 1958): 93–101; "The Waste of Fine Minds," *Life* 44 (April 7, 1958): 89–97; "Tryouts for Good Ideas," *Life* 44 (April 14, 1958): 117–25; and "Family Zest for Learning," *Life* 44 (April 21, 1958): 103–11.

41. "Crisis in Education," 33, 28.

42. Sloan Wilson, "It's Time to Close Our Carnival," *Life* 44 (March 24, 1958): 36–37.

43. For an overview of scientific manpower and its relation to education in particular, see Rudolph, *Scientists in the Classroom*, chaps. 2–3. On manpower generally, see, among many, David Kaiser, "Cold War Requisitions, Scientific Manpower, and the Production of American Physicists after World War II," *Historical Studies in the Physical Sciences* 33 (2002): 131–59; Cyrus C. M. Mody, "How I Learned to Stop Worrying and Love the Bomb, the Nuclear Reactor, the Computer, Ham Radio, and Recombinant DNA," *Historical Studies in the Natural Sciences* 38 (2008): 451–61; and Everett Mendelsohn, Merritt Roe Smith, and Peter Weingart, eds., *Science Technology and the Military* (Dordrecht: Kluwer Academic, 1988); and for contemporary analyses, Vannevar Bush, *Science: The Endless Frontier* (Washington, DC: Government Printing Office, 1945), and Derek J. de Solla Price, *Little Science, Big Science* (New York: Columbia University Press, 1963).

44. John R. Steelman, *Manpower for Research*, vol. 4 of *Science and Public Policy* (Washington, DC: Government Printing Office, 1947), 1, 64–66, 76, and appendix 2, pp. 48–149, submitted May 28, 1947.

45. National Science Foundation, *First Annual Report for the Fiscal Year Ended June 30, 1951* (Washington, DC: Government Printing Office, 1951), 17; Henry D. Smyth, "The Stockpiling and Rationing of Scientific Manpower," *Bulletin of the Atomic Scientists* 7 (February 1951): 38–42, 64; National Science Foundation, *Second Annual Report for the Fiscal Year Ended June 30, 1952* (Washington, DC: Government Printing Office, 1952), 75. For a history of the early National Science Foundation and governmental science policy, see Daniel Kevles, "The National Science Foundation and the Debate over Postwar Research Policy, 1942–1945: A Political Interpretation of *Science—The Endless Frontier*," *Isis* 68 (1977): 5–26, and Daniel Lee Kleinman, *Politics on the Endless Frontier: Postwar Research Policy in the United States* (Durham, NC: Duke University Press, 1995). For early NSF activity, see Milton Lomask, *A Minor Miracle: An Informal History of the National Science Foundation* (Washington, DC: National Science Foundation, 1976), and J. Merton England, *A Patron for Pure Science: The National Science Foundation's Formative Years, 1945–57* (Washington, DC: National Science Foundation, 1982), esp. pt. 3.

46. England, *Patron for Pure Science*, 217, has charts for fiscal years 1951–58; Congress, Senate, *Independent Offices Appropriations for FY1953*, 82nd Cong., 2nd sess., April 23, 1952, 327, and Congress, House, *Independent Offices Appropriations for FY 1953*, pt. 1, 82nd Cong., 2nd sess., January 15, 1952, 169–221.

47. England, *Patron for Pure Science*, 221. For the history and problematic nature of "basic," "fundamental," and "applied" research distinctions, see Shapin, *Scientific Life*; also Sabine Clarke, "Pure Science with a Practical Aim: The Meanings of Fundamental Research in Britain, circa 1916–1950," *Isis* 101 (2010): 285–311.

48. For a thorough history of the summer institute program, albeit one commissioned by the NSF, see Hillier Krieghbaum and Hugh Rawson, *An Investment in Knowledge: The First Dozen Years of the Science Foundation's Summer Institutes Programs to Improve Secondary School Science and Mathematics Teaching, 1954–1965* (New York: New York University Press, 1969), esp. 97, 102, 134–35, 160; also Rudolph, *Scientists in the Classroom*, 67–68. This first conference for high schools involved Prof. Carl Allendoerfer, later a participant in SMSG. SMSG later played a role in supplying these institutes with appropriate materials (see Krieghbaum and Rawson, *Investment in Knowledge*, 135).

49. Congress, House, *Independent Offices Appropriations for FY1956*, pt. 1, 84th Cong., 1st sess., February 9, 1955, 230–31; David Kaiser, "The Physics of Spin: Sputnik Politics and American Physicists in the 1950s," *Social Research* 73 (2006): 1225–52.

50. E.g., Congress, Senate, *Independent Offices Appropriations for FY1956*, 84th Cong., 1st sess., May 18, 1955, 426–28, and reprinted as Alan T. Waterman, "Role of the Federal Government in Science Education," *Scientific Monthly* 82 (June 1956): 286–93; see also England, *Patron for Pure Science*, 242.

51. Rudolph, *Scientists in the Classroom*, 219n24, 71–76.

52. Thomas: Congress, House, *Independent Offices Appropriations for FY1957*, pt. 1, 85th Cong., 1st sess., January 30, 1956, 522; Phillips: House, *Independent Offices Appropriations for FY1957*, 616; Krieghbaum and Rawson, *Investment in Knowledge*, chap. 11, and 203.

53. For example, the Special Interdepartmental Committee on the Training of Scientists and Engineers was formed in autumn 1953; Rudolph, *Scientists in the Classroom*, 63–64.

54. Langton T. Crane, *The National Science Foundation and Pre-College Science Education: 1950–1975*, report prepared for the Committee on Science and Technology, Congress, House, 94th Cong., 2nd sess. (Washington, DC: Government Printing Office, 1975), 42.

55. Rudolph, *Scientists in the Classroom*, 83–84; James R. Killian, Jr., *Sputnik, Scientists, and Eisenhower: A Memoir of the First Special Assistant to the President for Science and Technology* (Cambridge: MIT Press, 1977), 198; National Science Foundation, *Fifth Annual Report for the Fiscal Year Ended June 30, 1957* (Washington, DC: U.S. Government Printing Office, 1958), 73–74; House, Senate, *Independent Offices Appropriations for FY1958*, pt. 2, 85th Cong., 2nd sess., February 18, 1957, 1273.

56. Journalists traced the link reformers made between the launch and education, as evinced in such articles as Benjamin Fine, "Satellite Called Spur to Education: Soviet Success Shows Need for a Major U.S. Effort, College Heads Agree," *New York Times*, October 12, 1957, 3; Erwin Knoll, "Educators Told of School Need," *Washington Post*, October 12, 1957, C12; Benjamin Fine, "Soviet Satellite Blow to Educators in U.S.," *Boston Herald*, October 11, 1957, 22.

57. Mary Ann Callan, "Crisis Ahead in Education," *Los Angeles Times*, October 22, 1957, A1; see also her front-page articles on November 15, 17, and 18, 1957.

58. Zuoyue Wang, *In Sputnik's Shadow: The President's Science Advisory Committee and Cold War America* (New Brunswick, NJ: Rutgers University Press, 2008), 78–87, and chap. 5; Killian *Sputnik, Scientists, and Eisenhower*, 196. These appointments, coming after the launch of yet another Russian satellite in early November, responded to an increasing sense of urgency. For one early use of "missile czar," see Robert Donovan, "Killian Missile Czar: Ike Picks M.I.T. Head to Rush Research and Development," *Boston Daily Globe*, November 8, 1957, 1.

59. Killian, *Sputnik, Scientists, and Eisenhower*, chaps. 1–2; Dwight D. Eisenhower, *Waging War and Peace, 1956–1961*, vol. 2 of *The White House Years* (Garden City, NJ: Doubleday, 1965), chap. 8; Lyndon B. Johnson, *The Vantage Point: Perspectives of the Presidency, 1963–1969* (New York: Holt, Rinehart, and Winston, 1971), 270–78; for the "military-industrial complex": Dwight D. Eisenhower, "Farewell Radio and Television Address to the American People" [January 17, 1961], in Carroll W. Pursell, Jr., ed., *The Military-Industrial Complex*, 204–8 (New York: Harper and Row, 1972).

60. Z. Wang, *Sputnik's Shadow*, chap. 5; Equivalent 2013 dollars would be approximately $410 million and $1.09 billion, respectively. Funding data are located in the yearly *Annual Reports*. Kriegbaum and Rawson track exactly how quickly after the launch the NSF began to consider the options for requesting vastly increased amounts of money. See their *Investment in Knowledge*, esp. chap. 12; also Crane, *National Science Foundation*, 62.

61. Senate, *Science and Education*, 1170.

62. Killian, *Sputnik, Scientists, and Eisenhower*, 196; Congress, House, *Independent Offices Appropriations for FY1955*, pt. 1, 83rd Cong., 2nd sess., January 20, 1954, 753.

63. Wayne J. Urban, *More Than Science and Sputnik: The National Defense Education Act of 1958* (Tuscaloosa: University of Alabama Press, 2010); Clowse, *Brainpower for the Cold War*, 132–34. The provisions can be found in Congress, Senate, Committee on Labor and Public Welfare, *The National Defense Education Act of 1958*, 85th Cong., 2nd sess., September 5, 1958.

64. Congress, House, 85th Cong., 2nd sess., *Congressional Record* 104, pt. 15 (August 23, 1958): 19596. The tenuous link between national security and federal education legislation was widely mocked by conservatives during hearings; see, e.g., Strom Thurmond in the Senate and John James Flynt, Jr., in the House speaking against the bill, 85th Cong., 2nd sess., *Congressional Record* 104, pt. 13 (August 8 and 13,1958): 16570 and 17296.

65. For actual creation of SMSG: Congress, Senate, *Second Supplemental Appropriations Hearing*, 85th Cong., 2nd sess., February 26–28, 1958, 109; Na-

tional Science Foundation, *Seventh Annual Report for the Fiscal Year Ended June 30, 1957* (Washington, DC: Government Printing Office, 1957), 73–74. On other NSF course content programs, see Rudolph, *Scientists in the Classroom*; Richard J. Merrill and David W. Ridgway, *The CHEM Study Story* (San Francisco: W. H. Freeman, 1969); Suzanne Kay Quick, "Secondary Impacts of the Curriculum Reform Movement: A Longitudinal Study of the Incorporation of Innovations of the Curriculum Reform Movement into Commercially Developed Curriculum Programs" (Ph.D. diss., Stanford University, 1978).

66. Interview with Paul Rosenbloom, box 4RM19, R. L. Moore Legacy Collection (hereafter Moore Collection), 1890–1900, 1920–2009, Archives of American Mathematics, Dolph Briscoe Center for American History, University of Texas at Austin, 34–35; William Wooton, *SMSG: The Making of a Curriculum* (New Haven: Yale University Press, 1965), 9–11. Wooton's volume also has useful appendices which list participants in the various conferences and committees around this period.

67. Rosenbloom interview, box 4RM19, Moore Collection, 35.

68. Interview with Elsie Begle, box 4RM15, Moore Collection, 7; George Goodman, Jr., "Prof. Edward G. Begle, Chief Proponent of 'New Math,'" *New York Times*, March 3, 1978, B2.

69. Interview with Henry L. Pollak, transcript, box 4RN108, Moore Collection.

70. Rudolph, *Scientists in the Classroom*, esp. 79–80.

71. "Speeches—E. G. Begle," box 13.4/86–28/48, SMSGR; "List of Participants in the Work of SMSG, October 1966," in folder "List of Participants," box 13.5/86–28/68, SMSGR.

72. Wooton, *SMSG*, 41–42.

73. Rosenbloom interview, box 4RM19, Moore Collection, 57–63.

74. Mason: Congress, House, 85th Cong., 2nd sess., *Congressional Record* 104, pt. 3 (March 10, 1958): 3826; Bronk: Senate, *Science and Education*, 7 and 18.

75. Benjamin DeMott, "The Math Wars," *American Scholar* 31 (1962): 296–310, at 310.

76. For the more recent relationship of the federal government and education, see, for example, Kenneth W. Thompson, ed., *The Presidency and Education* (Lanham, MD: University Press of America, 1990); Hugh Davis Graham, *The Uncertain Triumph: Federal Education Policy in the Kennedy and Johnson Years* (Chapel Hill: University of North Carolina Press, 1984); Gareth Davies, *See Government Grow: Education Politics from Johnson to Reagan* (Lawrence: University Press of Kansas, 2007); Patrick J. McGuinn, *No Child Left Behind and the Transformation of Federal Education Policy, 1965–2005* (Lawrence: University Press of Kansas, 2006). President Barack Obama's administration, predictably, continued the trend of federal involvement in education with its Race to the Top initiative.

Chapter 3. The Textbook Subject

1. William Wooton, *SMSG: The Making of a Curriculum* (New Haven: Yale University Press, 1965), 14–15. Appendices of Wooton, *SMSG*, are helpful here. Some of the most centrally involved mathematicians were A. Adrian Albert (University of Chicago), Andy M. Gleason (Harvard), Albert E. Meder, Jr. (Rutgers), Edwin E. Moise (Michigan/Harvard), A. W. Tucker (Princeton), Marshall H. Stone (Chicago), and G. Baley Price (Kansas).

2. Thomas S. Kuhn, *The Structure of Scientific Revolutions*, 3rd ed. (Chicago: University of Chicago Press, 1996 [1962]), esp. 46, 137–38. Kuhn's claims continue to be supported by exemplary studies, most of which, however, focus on college-level textbooks. See, e.g., David Kaiser, ed., *Pedagogy and the Practice of Science: Historical and Contemporary Perspectives* (Cambridge: MIT Press, 2005); Kathryn Olesko, "Science Pedagogy as a Category of Historical Analysis: Past, Present, and Future," *Science and Education* 15 (2006): 863–80; Bernadette Bensaude-Vincent, "Textbooks on the Map of Science Studies," *Science and Education* 15 (2006): 667–70; Anders Lundgren and Bernadette Bensaude-Vincent, eds., *Communicating Chemistry: Textbooks and Their Audiences, 1789–1939* (Canton, MA: Science History Publications, 2000); Pere Grapí, "The Role of Chemistry Textbooks and Teaching Institutions in France at the Beginning of the Nineteenth Century in the Controversy about Berthollet's Chemical Affinities," in Peter Heering and Rolan Wittje, eds., *Learning by Doing: Experiments and Instruments in the History of Science Teaching*, 55–70 (Stuttgart: Franz Steiner Verlag, 2011); and an *Isis* Focus section edited by Marga Vicedo, titled "Textbooks in the Sciences," *Isis* 103 (2012).

3. Cf. Jean Lave, *Cognition in Practice: Mind, Mathematics, and Culture in Everyday Life* (Cambridge: Cambridge University Press, 1988).

4. For justifications of science curriculum in particular, see John Rudolph, *Scientists in the Classroom: The Cold War Reconstruction of American Science Education* (New York: Palgrave, 2002), esp. 5 and 176.

5. Alan R. Osborne and F. Joe Crosswhite, "Forces and Issues Related to Curriculum and Instruction, 7–12," in *A History of Mathematics Education in the United States and Canada, Thirty Second Yearbook*, 155–300 (Washington, DC: National Council of Teachers of Mathematics, 1970), 194–96; Karen Hunger Parshall and David E. Rowe, *The Emergence of the American Mathematical Research Community, 1876–1900: J. J. Sylvester, Felix Klein, and E. H. Moore* (Providence: American Mathematical Society, 1994), 401–19; also David Lindsay Roberts, *American Mathematicians as Educators, 1893–1923: Historical Roots of the "Math Wars"* (Chestnut Hill, MA: Docent Press, 2012).

6. Letter from John W. Green to Richard Brauer, March 7, 1958, and reply of March 10, 1958, in box 45, folder 54, American Mathematical Society Archives,

Brown University, Providence, RI. Alma Steingart kindly informed me of this letter.

7. For College Board Report's influence on SMSG, see "Speeches—E.G. Begle," box 4.1/86–28/48, School Mathematics Study Group Records, 1958–77 (hereafter SMSGR), Archives of American Mathematics, Dolph Briscoe Center for American History, University of Texas at Austin; SMSG, *Report of an Orientation Conference for Geometry with Coordinates, Chicago, Illinois, September 23, 1961* (Palo Alto, CA: Stanford University Press, 1962), 29; also E. E. Moise's report in "History and Philosophy of SMSG Writing Teams," box 13.7/86–28/81, SMSGR, 25.

8. College Entrance Examination Board, *Report of the Commission on Mathematics: Program for College Preparatory Mathematics* (New York: College Entrance Education Board, 1959), 5–6.

9. Nicholas Bourbaki, "The Architecture of Mathematics," *American Mathematical Monthly* 57 (1950): 221–32, at 230; the article was originally published as Nicolas Bourbaki, "L'architecture des mathématiques," in F. Le Lionnais, ed., *Les grands courants de la pensée mathématique*, 35–47 ([Marseille]: Cahiers du Sud, 1948), quotation at 45. For the relevance of the structural metaphor in a different setting, see Peter Galison, "Structure of Crystal, Bucket of Dust," in Apostolos Doxiadis and Barry Mazur, eds., *Circles Disturbed: The Interplay of Mathematics and Narrative*, 52–78 (Princeton: Princeton University Press, 2012).

10. On the overstatement: Leo Corry, *Modern Algebra and the Rise of Mathematical Structures* (Basel: BirkhäuserVerlag, 1996), esp. chap. 7. On Bourbaki generally, Maurice Mashaal, *Bourbaki: A Secret Society of Mathematicians*, trans. Anna Pierrehumbert (Providence: American Mathematical Society, 2006); David Aubin, "The Withering Immortality of Nicholas Bourbaki: A Cultural Connector at the Confluence of Mathematics, Structuralism, and the Oulipo in France," *Science in Context* 10 (1997): 297–342; Henri Cartan, "Nicholas Bourbaki and Contemporary Mathematics" [1958], *Mathematical Intelligencer* 2 (1980): 175–87; Joong Fang, *Bourbaki* (Hauppauge: Paideia Press, 1970); Liliane Beaulieu, "Bourbaki's Art of Memory," *Osiris*, 2nd ser., 14 (2002): 219–51.

11. See, e.g., Henri Cartan, Gustave Choquet, and Jacques Dixmier, *Structures algébriques et structures topologiques* (Geneva: Enseignement mathématique, 1958), a collection of essays presented at conferences for teachers of mathematics at the Henri-Poincaré Institute at the Sorbonne, in Paris, from February 1956 to June 1957.

12. Organization for European Economic Co-Operation, *New Thinking in School Mathematics* (1961), esp. 35–36.

13. SMSG, *Report of an Orientation Conference for Geometry with Coordinates*, 12–15; René Thom, "'Modern' Mathematics: An Educational and Philosophic Error," *American Scientist* 59 (1971): 695–99, at 695.

14. E. G. Begle, *Critical Variables in Mathematics Education: Findings from a Survey of the Empirical Literature* (Washington, DC: MAA and NCTM, 1979), 1; interview with Andrew Gleason, box 4RM17, R. L. Moore Legacy Collection (hereafter Moore Collection), 1890–1900, 1920–2009, Archives of American Mathematics, Dolph Briscoe Center for American History, University of Texas at Austin, 51–52; A. Adrian Albert, *Structure of Algebras* (New York: American Mathematical Society, 1939).

15. Marshall H. Stone, "Reminiscences of Mathematics at Chicago," in Peter Duren, ed., *A Century of Mathematics in America*, 2:183–90 (Providence: American Mathematical Society, 1989 [1976]), quotation at 185. See also Saunders Mac Lane, "Mathematics at the University of Chicago: A Brief History," 127–54, and Felix E. Browder, "The Stone Age of Mathematics on the Midway," 191–93, ibid.

16. Corry, *Modern Algebra*, esp. 289, 337, 355–57.

17. Marshall Stone, "The Revolution in Mathematics," *American Mathematical Monthly* 68 (1961): 715–34, at 715–17, 730, 733.

18. SMSG *Newsletter* 38 (August 1972), pt. 2.

19. See, e.g., the opening speech of Begle in folder "Speeches—E. G. Begle," box 13.4/86–28/48, and "Basic Principles," box 4.1/86–28/5, SMSGR.

20. SMSG, *Mathematics for Junior High School: Student's Text*, 2 vols., rev. ed. (New Haven: Yale University Press, 1961), 1:1.

21. Ibid., 1:8.

22. Ibid., 2:545, 578, 581–88. A number of these "unsolved" problems have indeed since been solved, but not by particularly young people, as it happened.

23. Ibid., 1:26.

24. Ibid., 1:37.

25. Ibid., 1:559. "Binary" in that there are two inputs (i.e., $2 \times 3 = 6$ represents a binary operation because integer multiplication uses two integers to produce a third; other operations, such as finding the square root of a number, are not binary).

26. Ibid., 2:38. In the interest of simplicity, I have slightly modified the equations; the text uses raised negatives for each negative number, to emphasize it as a symbol distinct from the operation of subtraction. While such choices are relevant to understanding the order of presentation of the text, they do not change the larger point of the demonstration.

27. Congress, Senate, Committee on Labor and Public Welfare, *Science and Education for National Defense*, 85th Cong., 2nd sess., January 21, 1958, 6–7; National Science Foundation, *Eighth Annual Report for Year Ended June 30, 1958* (Washington, DC: Government Printing Office, 1958), 64–66; Stone, as quoted in Amy Dahan Dalmedico, "An Image Conflict in Mathematics after 1945," in Umberto Bottazzini and Amy Dahan Dalmedico, eds., *Changing Images in Mathematics: From the French Revolution to the New Millennium*, 223–53 (New York: Taylor and Francis, 2001), at 232.

28. Warren Colburn, *First Lessons in Arithmetic, on the Method of Pestalozzi, with Some Improvements* (Boston: Cummings, Hilliard, 1825). Nicolas Pike, *New and Complete System of Arithmetick, Composed for Use of Citizens of the United States*, 7th ed., ed. Nathanial Lord (Boston: Thomas and Andrews, 1809). More generally, see Patricia Cline Cohen, *A Calculating People: The Spread of Numeracy in Early America* (Chicago: University of Chicago Press, 1982), esp. chaps. 1–4.

29. "Improvement in Arithmetical Instruction," *American Journal of Education* 2 (1827): 30–37, at 30 (emphasis in original).

30. The examples in this paragraph are discussed in Patricia Cline Cohen, "Numeracy in Nineteenth-Century America," in George M. A. Stanic and Jeremy Kilpatrick, eds., *A History of School Mathematics*, 1:43–76 (Reston, VA: National Council of Teachers of Mathematics, 2003), 62–65.

31. Herbert M. Kliebard and Barry M. Franklin, "The Ascendance of Practical and Vocational Mathematics, 1893–1945: Academic Mathematics under Siege," in Stanic and Kilpatrick, *History of School Mathematics*, 1:399–440.

32. Commission on the Secondary School Curriculum of the Progressive Education Association, *Mathematics in General Education: A Report of the Committee on the Function of Mathematics in General Education* (New York: D. Appleton-Century, 1940), esp. 8, 11–12 (emphasis added). A focus on practicality was hardly limited to mathematics and was a theme in the transformation of science education more broadly in this period. For the case of physics, see, e.g., David Donahue, "Serving Students, Science, or Society? The Secondary School Physics Curriculum in the United States, 1930–1965," *History of Education Quarterly* 33 (1993): 321–52.

33. National Committee on Mathematical Requirements, *The Reorganization of Mathematics in Secondary Education* ([Oberlin, OH]: MAA, 1923), as discussed in Roberts, *American Mathematicians as Educators*; Virgil S. Mallory, *Mathematics for Victory: An Emergency Course* (Chicago: Benj. H. Sanborn, 1943); George Pólya, *How to Solve It: A New Aspect of Mathematical Method* (Princeton: Princeton University Press, 1945), and G. Pólya, *Induction and Analogy in Mathematics*, 2 vols. (Princeton: Princeton University Press, 1954). Although Begle's wife later said SMSG's leader admired Pólya greatly, Pólya did not seem to appreciate SMSG's textbooks: interview with Elsie Begle, transcript, box 4RM15, Moore Collection, 15.

34. SMSG's textbooks for more advanced material than algebra and geometry did not sell well, in part because relatively few students took four years of mathematics in high school. SMSG promoted more advanced courses but the organization did not last long enough to see the fruits of this particular effort. As students gradually took more mathematics in high school over the intervening years, calculus and trigonometry courses have, of course, proliferated.

35. SMSG, *Mathematics for High School: Geometry: Teacher's Commentary*,

rev. ed. (New Haven: Yale University Press, 1961), 12 (emphasis in original); also SMSG, *Mathematics for High School: Geometry*, rev. ed. (New Haven: Yale University Press, 1961), chap. 2, and p. c of the appendices.

36. SMSG, *Geometry*, 339–340. They were not claiming that this was a *new* proof, but rather one that fit with SMSG's overall geometric program. For the Euclidean original, Thomas L. Heath, *The Thirteen Books of Euclid's Elements: Translated from the Text of Heiberg*, 2nd rev. ed., 3 vols. (Cambridge: Cambridge University Press, 1926), 1:349–50.

37. Letter from Robert J. Walker to Edward Begle, April 5, 1960, folder "New Geometry," box 13.5/86–28/60, SMSGR.

38. Namely, the Jordan Curve theorem; see Morris Kline, *Mathematical Thought from Ancient to Modern Times*, 3 vols. (New York: Oxford University Press, 1972), 3:1017; Jeremy Gray, *Plato's Ghost: The Modernist Transformation of Mathematics* (Princeton: Princeton University Press, 2008), 229; Ivor Grattan-Guinness, *The Rainbow of Mathematics: A History of the Mathematical Sciences* (New York: Norton, 1997), 692. On non-Euclidean geometry and incontrovertible spatial truths, see Joan L. Richards, *Mathematical Visions: The Pursuit of Geometry in Victorian England* (Boston: Academic Press, 1988), and Jeremy Gray, *János Bolyai, Non-Euclidean Geometry, and the Nature of Space* (Cambridge: MIT Press, 2004).

39. Gray, *Plato's Ghost*, 271, 400, 456. The introduction of such varied and controversial objects has proven fruitful for both philosophical and historical examinations of mathematics. See Imre Lakatos, *Proofs and Refutations: The Logic of Mathematical Discovery*, ed. John Worrall and Elie Zahar (Cambridge: Cambridge University Press, 1976), and Gray, *Plato's Ghost*, 235–39.

40. Oswald Veblen, "A System of Axioms for Geometry," *Transactions of the American Mathematical Society* 5 (1904): 343–84, at 343; the Hilbert quotation is related, among many other places, in Stewart Shapiro, *Philosophy of Mathematics: Structure and Ontology* (New York: Oxford University Press, 1997), 157. Hilbert published his own volume in 1899 attempting to lay a new foundation for geometry: David Hilbert, *Foundations of Geometry*, trans. Leo Unger (La Salle, IL: Open Court, 1971 [from material developed in 1898–99]); George D. Birkhoff, "A Set of Postulates for Plane Geometry, Based on Scale and Protractor," *Annals of Mathematics*, 2nd ser., 33, no. 2 (1932): 329–45; quotation from George David Birkhoff and Ralph Beatley, *Geometry*, prelim. ed. (Boston: Spaulding-Moss, 1933), i.

41. SMSG, *Mathematics for High School: First Course in Algebra: Teacher's Commentary*, rev. ed. (New Haven: Yale University Press, 1961), x, 202–3, and 341–43; SMSG, *Mathematics for High School: First Course in Algebra: Student's Text*, rev. ed. (New Haven: Yale University Press, 1961), 313, 347, and 352. For the early history of algebra, see, e.g., Kline, *Mathematical Thought from Ancient to Modern Times*, esp. 1:191–95, and for the history of early symbols, the clas-

sic, Florian Cajori, *A History of Mathematical Notations*, 2 vols. (Chicago: Open Court, 1928).

42. SMSG, *Algebra: Student's Text*, 135–37.

43. SMSG, *Algebra: Teacher's Commentary*, 343.

44. College Entrance Examination Board, *Report of the Commission on Mathematics*, 22; SMSG, *Mathematics for Junior High School: Teacher's Commentary*, rev. ed., 2 vols. (New Haven: Yale University Press, 1961), 1:381.

45. George F. Carrier et al., "Applied Mathematics: What Is Needed in Research and Education: A Symposium," *SIAM Review* 4 (1962): 297–320, at 300.

46. Ibid., 298–99. Courant had espoused similar views many years earlier in the opening pages to his textbook: Richard Courant and Herbert Robbins, *What Is Mathematics?* (New York: Oxford University Press, 1941).

47. Carrier et al., "Applied Mathematics," 303–4. The original text misspells "Friedman" as "Freedman."

48. SMSG, *Mathematics for Junior High School: Teacher's Commentary*, 1:381–82.

49. H. P. Greenspan, "Applied Mathematics as a Science," *American Mathematical Monthly* 68 (1961): 872–80; Carrier et al., "Applied Mathematics," 297–301 and 310–11; on institutional arrangements, Amy Dahan Dalmedico, "L'essor des mathématiques appliquées aux États-Unis," *Revue d'Histoire des Mathématiques* 2 (1996): 149–213, esp. 189–98; see also Peter Lax, "The Flowering of Applied Mathematics in America," in Duren, *Century of Mathematics in America*, 2:455–66; these trends, and tensions, are evident in Committee on Support of Research in the Mathematical Sciences, *The Mathematical Sciences: A Report* (Washington, DC: National Academy of Sciences, 1968).

50. Morris Kline, "Math Teaching Reforms Assailed as Peril to U.S. Scientific Progress," *New York University Alumni News* (October 1961): 1, 3, 8; see also the compilation of his views in M. Kline, *Why Johnny Can't Add: The Failure of the New Math* (New York: St. Martin's Press, 1973). Kline had been a critic of mathematics education even prior to SMSG, e.g., M. Kline, "Mathematics Texts and Teachers: A Tirade," *Mathematics Teacher* 49 (March 1956): 162–72.

51. "On the Mathematics Curriculum of the High School," *American Mathematical Monthly* 69 (March 1962): 189–93, esp. 191; Robert C. Toth, "Teaching of the 'New Math' Stirs Wide Debate among Teachers," *New York Times*, September 21, 1962, 31; for context, see David Lindsay Roberts, "The BKPS Letter of 1962: The History of a 'New Math' Episode," *Notices of the AMS* 51 (2004): 1062–63.

52. E. G. Begle, "Remarks on the Memorandum 'On the Mathematics Curriculum of the High School,'" *American Mathematical Monthly* 69 (May 1962): 425–26; see also Roberts, "BKPS Letter of 1962," as well as the folder "Re: Bers-Kline . . . ," box 13.5/86–28/53, SMSGR.

53. Letter from Lipman Bers to E. G. Begle, December 28, 1959, in folder "Advisory Committee," box 4.1/86–28/1, SMSGR.

54. "On the Mathematics Curriculum of the High School," 190; René Thom, "Modern Mathematics: Does It Exist?" in A. G. Howson, ed., *Developments in Mathematical Education: Proceedings of the Second International Conference on Mathematical Education*, 194–209 (Cambridge: Cambridge University Press, 1973), 206; Kline's historical research culminated in *Mathematical Thought from Ancient to Modern Times*.

55. Begle, "Remarks on the Memorandum," 426; letter from Marshall Stone to E. G. Begle, November 20, 1961, in "Re: Bers-Kline . . . ," box 13.5/86–28/53, SMSGR.

56. Kline, "Math Teaching Reforms," 8; for Courant's influence: G. L. Alexanderson, "An Interview with Morris Kline: Part 2," *Two-Year College Mathematics Journal* 10 (1979): 259–64, at 262.

57. Kline, *Why Johnny Can't Add*, 91–92; Morris Kline, "Logic versus Pedagogy," *American Mathematical Monthly* 77 (March 1970): 264–82, at 279; Glenn James, ed., *Tree of Mathematics* (Pacoima, CA: Digest Press, 1957). The "tree of mathematics" builds on a long-standing use of treelike structures to represent divisions of knowledge, e.g., John E. Murdoch, *Album of Science: Antiquity and the Middle Ages* (New York: Charles Scribner's Sons, 1984), esp. 38–51.

58. G. Baley Price, "Mathematics and Education," in Lee C. Deighton, ed., *The Encyclopedia of Education*, 6:81–90 (New York: Macmillan, 1971), 86; for the comparison of queen and handmaiden, see Eric Temple Bell, *Mathematics: Queen and Servant of Science* (New York: McGraw-Hill, 1951).

59. Albert E. Meder, Jr., "The Ancients versus the Moderns—A Reply," *Mathematics Teacher* 51 (October 1958): 428–33, at 433.

60. Kline, "Math Teaching Reforms," 8.

Chapter 4: The Subject in Itself

1. Arthur Bestor, *Educational Wastelands: The Retreat from Learning in our Public Schools*, 2nd ed. (Urbana: University of Illinois Press, 1985 [1953]), 24 and 59.

2. George Orwell, *Nineteen Eighty-Four*, ed. Bernard Crick (Oxford: Clarendon Press, 1984 [1949]), 226, 374–75.

3. "Third Draft of SMSG Position Statement," June 29, 1962, box 13.4/86–28/45, School Mathematics Study Group Records, 1958–77 (hereafter SMSGR), Archives of American Mathematics, Dolph Briscoe Center for American History, University of Texas at Austin, 6.

4. Jeremy J. Gray, *Plato's Ghost: The Modernist Transformation of Mathematics* (Princeton: Princeton University Press, 2008), 134, 159–70; Ivor Grattan-Guinness, *The Search for Mathematical Roots, 1870–1940: Logics, Set Theories and the Foundation of Mathematics from Cantor through Russell to Gödel*

(Princeton: Princeton University Press, 2000), esp. chaps. 3–4; José Ferreirós, *Labyrinth of Thought: A History of Set Theory and Its Role in Modern Mathematics* (Basel: Birkhäuser Verlag, 1999). Hilbert, as quoted in Ferreirós, *Labyrinth of Thought*, 149, who cites Leo Corry, *David Hilbert and the Axiomatization of Physics (1898–1918): From "Grundlagen der Geometrie" to "Grundlagen der Physik"* (Dordrecht: Kluwer Academic, 2004), 379. Hilbert also addressed the philosophical trends in mathematics in his list of "problems" in 1900; see Jeremy J. Gray, *The Hilbert Challenge* (New York: Oxford University Press, 2000).

5. On the use of textbooks, see Jeanne Moulton, *How Do Teachers Use Textbooks? A Review of the Research Literature* (Washington, DC: U.S. Agency for International Development, 1997). "Secret garden" is a phrase from Sir David Eccles in a slightly different sense; see Gary McCulloch, "The Politics of the Secret Garden: Teachers and the School Curriculum in England and Wales," in Christopher Day et al., eds., *The Life and Work of Teachers: International Perspectives in Changing Times*, 26–37 (London: Falmer Press, 1999). On the gap between midcentury curriculum designers' intentions and reality in a different context, see John Rudolph, "Teaching Materials and the Fate of Dynamic Biology in American Classrooms after Sputnik," *Technology and Culture* 53 (2012): 1–36.

6. There is a growing literature putting the body back into the mathematical and technical. See, e.g., Hélène Mialet, *Hawking Incorporated: Stephen Hawking and the Anthropology of the Knowing Subject* (Chicago: University of Chicago Press, 2012); Alison Adam, *Artificial Knowing: Gender and the Thinking Machine* (London: Routledge, 1998); and Steven Shapin, "The Philosopher and the Chicken: On the Dietetics of Disembodied Knowledge," in Christopher Lawrence and Steven Shapin, eds., *Science Incarnate: Historical Embodiments of Natural Knowledge*, 21–50 (Chicago: University of Chicago Press, 1998).

7. Howard F. Fehr, "Modern Mathematics and Good Pedagogy," *Arithmetic Teacher* 10 (November 1963): 402–11, at 403–4.

8. Milgram, as quoted in Thomas Blass, *The Man Who Shocked the World: The Life and Legacy of Stanley Milgram* (New York: Basic Books, 2004), 100; see also Rebecca Lemov, *World as Laboratory: Experiments with Mice, Mazes, and Men* (New York: Hill and Wang, 2005), 222–28; T. W. Adorno et al., *The Authoritarian Personality* (New York: Harper, 1950); Ellen Herman, *Romance of American Psychology: Political Culture in the Age of Experts* (Berkeley: University of California Press, 1995), 58–60; Jamie Cohen-Cole, *The Open Mind: Cold War Politics and the Sciences of Human Nature* (Chicago: University of Chicago Press, 2014).

9. Susan L. Carruthers, *Cold War Captives: Imprisonment, Escape, and Brainwashing* (Berkeley: University of California Press, 2009), chap. 5, esp. 207–13; Ron Robin, *The Making of the Cold War Enemy* (Princeton: Princeton University Press, 2001), chap. 8, esp. 164.

10. Mark Solovey and Hamilton Cravens, eds., *Cold War Social Science: Knowledge Production, Liberal Democracy, and Human Nature* (New York: Palgrave Macmillan, 2012); Jamie Cohen-Cole, "The Creative American: Cold War Salons, Social Science, and the Cure for Modern Society," *Isis* 100 (2009): 219–62, and J. Cohen-Cole, "Thinking about Thinking in Cold War America" (Ph.D. diss., Princeton University, 2003), esp. chap. 4; also Robin, *Making of the Cold War Enemy*; Margaret Peacock, *Cold War Kids: Images of Soviet-American Childhood and the Collapse of Consensus, 1945–1968* (Chapel Hill: University of North Carolina Press, forthcoming); and Herman, *Romance of American Psychology*, esp. chap. 3.

11. See, e.g., Robert K. Merton, "Science and Democratic Social Structure" [1942], in R. Merton, *Social Theory and Social Structure*, rev. ed., 550–61 (Glencoe, IL: Free Press, 1957); Thomas S. Kuhn, *The Structure of Scientific Revolutions* (Chicago: University of Chicago Press, 1962); David A. Hollinger, "Science as Weapon in *Kulturekämpfe* in the United States during and after World War II," in D. Hollinger, *Science, Jews, and Secular Culture: Studies in Mid-Twentieth-Century American Intellectual History*, 155–74 (Princeton: Princeton University Press, 1996).

12. Michael Polanyi, *Science, Faith and Society* (London: Oxford University Press, 1946), 58. For the milieu of Polanyi in particular, see Mary Jo Nye, "Historical Sources of Science-as-Social-Practice: Michael Polanyi's Berlin," *Historical Studies in the Physical and Biological Sciences* 37 (2007): 409–34.

13. Ludwig Wittgenstein, *Philosophical Investigations*, trans. G. E. M. Anscombe (Oxford: Basil Blackwell, 1968), as well as Wittgenstein, *Remarks on the Foundations of Mathematics*, rev. ed., ed. G. H. Von Wright, R. Rhees, and G. E. M. Anscombe, and trans. G. E. M. Anscombe (Cambridge: MIT Press, 1978), and Wittgenstein, *Lectures on the Foundations of Mathematics, Cambridge 1939*, ed. Cora Diamond (Chicago: University of Chicago Press, 1976). My analysis of his examples relies particularly on Saul A. Kripke, *Wittgenstein on Rules and Private Language: An Elementary Exposition* (Oxford: Blackwell, 1982), and David Bloor, *Wittgenstein, Rules and Institutions* (London: Routledge, 1997).

14. This point was made in David Bloor, "Wittgenstein and Mannheim on the Sociology of Mathematics," *Studies in History and Philosophy of Science* 4 (1973): 173–91, at 181; cf. David Kaiser, "A Mannheim for All Seasons: Bloor, Merton, and the Roots of the Sociology of Scientific Knowledge," *Science in Context* 11 (1998): 51–87.

15. H. M. Collins, *Changing Order: Replication and Induction in Scientific Practice*, 2nd ed. (Chicago: University of Chicago Press, 1992).

16. Cambridge Conference on School Mathematics, *Goals for School Mathematics* (Boston: Houghton Mifflin, 1963), 35.

17. "NSF General 1958–1959," box 4.1/86–28/4, SMSGR.

18. Letter from Begle to Marshall Stone, July 23, 1958, in folder "Primary Schools," box 13.4/86–13/4, SMSGR; SMSG, *Report of a Conference on Elementary School Mathematics* (New Haven: Yale University Press, 1959); "Advisory Committee," box 4.1/86–28/1, SMSGR.

19. See folder "Conference on Future Resp. of School Math," box 13.5/86–28/53, SMSGR. While financial documentation is scattered throughout the archives, see "Budget Proposals" in box 13.6/86–28/71, SMSGR, for the $8.3 million award given on June 30, 1961; as well as William Wooton, *SMSG: The Making of a Curriculum* (New Haven: Yale University Press, 1965), 107.

20. Folders "Elementary Panel" and "Elementary School 1964—Inactive," box 13.4/86–28/45; the "SMSG Gross Sales Report" in folders "Advisory Committee—Replies," box 4.1/86–28/1; "SMSG Financial Reports," box 13.5/86–28/61; and the "Recommendations to the Advisory Committee of SMSG on Elementary School Mathematics," in folder "Elementary School Mathematics Ad Hoc Comm Feb 59," box 13.5/86–28/54, SMSGR.

21. The following account is taken from SMSG, *Mathematics for the Elementary School, Book 1 Unit 53: Teacher's Commentary* (Palo Alto: Stanford University Press, 1965), and SMSG, *Mathematics for the Elementary School, Book 1* (Palo Alto: Stanford University Press, 1965). The authors are listed as Leslie Beatty of the Chula Vista (Calif.) City School District, E. Glenadine Gibb of the State (Teachers') College of Iowa, William T. Guy of the University of Texas, Stanley B. Jackson of the University of Maryland, Irene Sauble of the Detroit Public Schools, Marshall H. Stone of the University of Chicago, J. Fred Weaver of Boston University, and Raymond L. Wilder of the University of Michigan. This volume had been revised in summer 1965 based on testing conducted the previous school year.

22. David Roberts, "Mathematicians in the Schools: The 'New Math' as an Arena of Professional Struggle, 1950–1970," unpublished talk given at the History of Science Annual Meeting, November 20, 2004, 12; see also the interview with James Wilson and Jeremy Kilpatrick, box 4RM17, R. L. Moore Legacy Collection, 1890–1900, 1920–2009, Archives of American Mathematics (hereafter Moore Collection), Dolph Briscoe Center for American History, University of Texas at Austin, 15–16.

23. For a contemporaneous statement of this position: E. Glenadine Gibb, Phillip S. Jones, and Charlotte W. Junge, "Number and Operation," in *The Growth of Mathematical Ideas: Grades K–12, Twenty-fourth Yearbook of NCTM*, 7–64 (Washington, DC: National Council of Teachers of Mathematics, 1959), 8.

24. SMSG, *Report of a Conference on Elementary School Mathematics*, 7.

25. SMSG, *Studies in Mathematics*, vol. 13: *Inservice Course in Mathematics for Primary School Teachers*, rev. ed. (Palo Alto: Stanford University Press, 1965–66), 307.

26. Technically, it is the union of two "disjoint" sets—sets that have no members in common—a point made in the teacher's commentary but not in the student's edition.

27. Minutes from Advisory Committee meeting, April 9, 1960, folder "Drafts—Old Minutes," box 4.1/86–28/1, SMSGR.

28. James A. McLellan and John Dewey, *The Psychology of Number and Its Applications to Methods of Teaching Arithmetic* (New York: D. Appleton, 1895), esp. 86; John Dewey, *How We Think* (Boston: D. C. Heath, 1910), 61. Dewey hoped that his work in mathematics might enable him to integrate the "Hegelian logic of quantity" into psychology and then use McLellan's pedagogical methods to frame the concept for teachers. These efforts were some of Dewey's only collaborative writings, and the young professor hoped to gain financially from them. See Jay Martin, *The Education of John Dewey: A Biography* (New York: Columbia University Press, 2002), 110–11, 191.

29. Letters in folder "Chicago Conference: General," box 4.1/86–28/14, SMSGR; list from Wooton, *SMSG*, 159.

30. Marshall Stone, "Fundamental Issues in the Teaching of Elementary School Mathematics," in SMSG, *Report of a Conference on Elementary School Mathematics*, 7–8. On the lack of consensus, see Francis G. Lankford, Jr., "Implications of the Psychology of Learning for the Teaching of Mathematics," in *Growth of Mathematical Ideas*, 405–30; and on the field's heterogeneity, perhaps resulting from psychology's rapid growth, see James H. Capshew, *Psychologists on the March: Science, Practice, and Professional Identity in America, 1929–1969* (Cambridge: Cambridge University Press, 1999). Although Begle wrote very little on educational psychology directly, one of his close associates—Jeremy Kilpatrick—noted much later that Begle was skeptical of just about everything psychologists did. See interview with Wilson and Kilpatrick, box 4MR17, Moore Collection, 12.

31. The conference proceedings were published as Jerome Bruner, *The Process of Education* (New York: Vintage Books 1963 [1960]). On the conference, see John Rudolph, "From World War to Woods Hole: The Impact of Wartime Research Models on Curriculum Reform," *Teachers College Record* 104 (2002): 212–41.

32. Bruner, *Process of Education*, 6. There was in fact little agreement among psychologists about whether "transfer of training" actually occurred. See National Academy of Sciences—National Research Council, "Psychological Research in Education: Report of a Conference Sponsored by the Advisory Board on Education" (Washington, DC: NAS-NRC, 1958), 13–14.

33. Bruner, *Process of Education*, 7 and 12.

34. Jean Piaget, "Comments on Mathematical Education," in A. G. Howson, ed., *Developments in Mathematical Education: Proceedings of the Second In-*

ternational Conference on Mathematical Education, 79–87 (Cambridge: Cambridge University Press, 1973), at 79.

35. Jean Piaget, "Science of Education and the Psychology of the Child" [1965], in Howard E. Gruber and J. Jacques Vonèche, eds., *The Essential Piaget*, 695–725 (London: Routledge and Kegan Paul, 1977), at 702. See also Charles J. Brainerd, *Piaget's Theory of Intelligence* (Englewood Cliffs, NJ: Prentice-Hall, 1978).

36. Bruner, *Process of Education*, 44, 33; cf. Jean Piaget, "How Children Form Mathematical Concepts," *Scientific American* 189, no. 5 (November 1953): 74–79.

37. Paul C. Rosenbloom, "Mathematics in the World of Children," in Thomas L. Saaty and F. Joachim Weyl, eds., *The Spirit and the Uses of the Mathematical Sciences*, 68–104 (New York: McGraw-Hill, 1969), esp. 81–84; and "Elementary School Panel Minutes of June 20, 1964," in folder "Elementary School, 1964—Inactive," box 13.4/86–28/45, SMSGR.

38. See P. C. Rosenbloom, "Teaching Mathematics to Gifted Students," and Newton S. Hawley and Patrick Suppes, "Geometry in the First Grade," in SMSG, *Conference on Elementary School Mathematics*, 33–35 and 36–37, respectively.

39. William A. Brownell and Harold E. Moser, *Meaningful vs. Mechanical Learning: A Study in Grade III Subtraction* (Durham, NC: Duke University Press, 1949), esp. 155–57. One textbook in "meaningful arithmetic" based on Brownell's work was Francis J. Mueller, *Arithmetic: Its Structure and Concepts* (Englewood Cliffs, NJ: Prentice-Hall, 1956); Mueller himself was careful to distinguish "meaningful arithmetic" from the new math on the basis of the involvement of the professional mathematician, e.g., F. Mueller, *Understanding the New Elementary School Mathematics: A Parent–Teacher Guide* (Belmont, CA: Dickenson Publishing, 1965), 4–5.

40. For a brief overview of his influence, see Jeremy Kilpatrick and J. Fred Weaver, "The Place of William A. Brownell in Mathematics Education," *Journal for Research in Mathematics Education* 8 (November 1977): 382–84. Interestingly, this article was written by two protégés of Begle, who were each closely involved with SMSG; they later claimed that Brownell, with his careful focus on scientific examination, was an important influence on Begle: interview with Wilson and Kilpatrick, box 4RM17, Moore Collection, 47.

41. William Brownell, "Principles in Curriculum Construction," in SMSG, *Conference on Elementary School Mathematics*, 15–17. Brownell's gloss on the history of educational psychology was a standard one, e.g., the nearly identical (but far more detailed) summary given in James A. Bell, "Trends in Secondary Mathematics in Relation to Psychological Theories: 1893–1970" (Ph.D. diss., University of Oklahoma, 1971), although Bell was apparently unaware of Brownell's speech.

42. Quotation from interview with Wilson and Kilpatrick, box 4RM17, Moore Collection, 19; also Bell, "Trends in Secondary Mathematics," 130–32.

43. SMSG, *Mathematics for High School: Geometry: Teacher's Commentary*, rev. ed. (New Haven: Yale University Press, 1961), 515–16. Although this is not from an elementary text, it is consistent with SMSG's overall approach; the main author of the geometry series was Edwin Moise, a central SMSG writer.

44. Ibid., 516, 518–20.

45. Constance H. Tolkan, "A Parent Looks at the New Math," *Independent School Bulletin* 31 (October 1971): 53–55, at 53.

46. Indeed, Begle's widow said plainly about his attitude toward students, "Ed was an egalitarian." Interview with Elsie Begle, box 4RM15, Moore Collection, 20.

47. SMSG, *Inservice Course in Mathematics*, chap. 1, esp. pp. 3–9; for its wider contemporary resonance, see, e.g., Basil B. Bernstein, *Class, Codes, and Control*, 3 vols. (London: Routledge and Kegan Paul, 1971).

48. SMSG, *Inservice Course in Mathematics*, 316–21; similar "special editions" were available for nearly all the published SMSG materials.

49. National Council of Teachers of Mathematics' Regional Orientation Conferences in Mathematics (NCTM), *The Revolution in School Mathematics: A Challenge for Administrators and Teachers* (Washington, DC: NCTM, 1961), 1–5; see also interview with Rosenbloom, box 4RM19, Moore Collection, 79. For the origin of the "head and hand" distinction, see Steven Shapin and Barry Barnes, "Head and Hand: Rhetorical Resources in British Pedagogical Writing, 1770–1850," *Oxford Review of Education* 2 (1976): 231–54; Raymond Williams, *Keywords: A Vocabulary of Culture and Society*, new ed. (New York: Oxford University Press, 1983 [1976]), 201–2. This distinction remained a common refrain in the new math era, e.g., H. G. Rickover, *Education and Freedom* (New York: Dutton, 1959), 17–19.

50. E.g., reforms in physics: John Rudolph, *Scientists in the Classroom: The Cold War Reconstruction of American Science Education* (New York: Palgrave, 2002); Jerrold Zacharias of the Physical Sciences Study Committee was later involved with Educational Services, Inc., and the Educational Development Center, which indeed produced elementary materials using National Science Foundation money. David Kaiser has examined changes in the public perception of physics at the university level: David Kaiser, "The Postwar Suburbanization of American Physics," *American Quarterly* 56 (2004): 851–88, and D. Kaiser, *How the Hippies Saved Physics: Science, Counterculture, and the Quantum Revival* (New York: W. W. Norton, 2011).

51. SMSG, *Mathematics for Junior High School: Student's Text*, rev. ed., 2 vols. (New Haven: Yale University Press, 1961), 1:613.

52. SMSG, *Mathematics for Junior High School: Teacher's Commentary*, rev. ed., 2 vols. (New Haven: Yale University Press, 1961), 1:311–12.

Chapter 5. The Subject in the Classroom

1. This might be termed the distinction between "policy talk" and "policy action": David Tyack and Larry Cuban, *Tinkering toward Utopia: A Century of Public School Reform* (Cambridge: Harvard University Press, 1995), chap. 2.

2. Hillier Krieghbaum and Hugh Rawson, *An Investment in Knowledge: The First Dozen Years of the Science Foundation's Summer Institutes Programs to Improve Secondary School Science and Mathematics Teaching, 1954–1965* (New York: New York University Press, 1969), 316.

3. From speech in "Speeches—E. G. Begle," box 13.4/86–28/48, in School Mathematics Study Group Records, 1958–77 (hereafter SMSGR), Archives of American Mathematics, Dolph Briscoe Center for American History, University of Texas at Austin.

4. E. G. Begle, "The School Mathematics Study Group," *Mathematics Teacher* 51 (December 1958): 618.

5. SMSG, *Report of an Orientation Conference for Geometry with Coordinates, Chicago, Illinois, September 23, 1961* (Palo Alto: Stanford University Press, 1962), 91.

6. E. P. Northrop, "Modern Mathematics and the Secondary School Curriculum," *Mathematics Teacher* 48 (October 1955): 386–93, at 387 and 390; Albert E. Meder, Jr., "Modern Mathematics and Its Place in the Secondary School," *Mathematics Teacher* 50 (October 1957): 418–23, at 423; Howard F. Fehr, "Helping Our Teachers," *Mathematics Teacher* 50 (October 1957): 451–52.

7. National Council of Teachers of Mathematics' Regional Orientation Conferences in Mathematics (NCTM), *The Revolution in School Mathematics: A Challenge for Administrators and Teachers* (Washington, DC: NCTM, 1961).

8. Numbers from "List of Participants in the Work of the School Mathematics Study Group," October 1966, box 13.6/86–28/68, SMSGR. While the biology curriculum project was slightly more expensive overall, much of that cost was actually in implementation rather than development, given that the biology project published actual textbooks for use (rather than simply models). See National Science Foundation, Science Curriculum Review Team, *Pre-College Science Curriculum Activities of the National Science Foundation* (Washington, DC: National Science Foundation, 1975), 2:168. Rosenbloom's comments in George F. Carrier, R. Courant, Paul Rosenbloom, and C. N. Yang, "Applied Mathematics: What Is Needed in Research and Education, A Symposium," *SIAM Review* 4 (1962): 297–320, at 314–16.

9. "What Makes Them Good," *Time*, October 21, 1957, 52; Newton School Committee, *Proceedings*, Newton Public School Archives, Newton, MA, October 21, 1957, 184; and March 5, 1958, 296; Floyd Rinker, "Subject Matter, Students, Teachers, Methods of Teaching, and Space Are Redeployed in the

Newton, Massachusetts, High School," *Bulletin of the National Association of Secondary-School Principals* 42 (January 1958): 69–80.

10. Newton School Committee, *Proceedings*, March 10, 1958, 300; February 24, 1958, 284; May 19, 1958, 351; and December 23, 1957, 233, all in Newton Public School Archives.

11. W. Eugene Ferguson, "Mathematics in Newton," *NAASP Bulletin* 52 (April 1968): 55–64; Newton Public Schools, "Math—A Stimulant," *Annual Report* 118 (1957–58): 44–45; "Mathematics—Experimentation Continues," *Annual Report* 119 (1958–59): 7–10; "Mathematics Instruction in the Modern Manner: New Approaches to the Teaching of Mathematics Are Being Used on the Secondary Level," *Annual Report* 120 (1959–60): 80–83; all *Annual Reports* are in Newton Public School Archives.

12. See, for example, "Applications" folder, box 4.1/86–28/5, SMSGR.

13. "Colorado" folder, box 4.1/86–28/11, SMSGR.

14. Ibid.

15. Wendell Hall, "Eugene Develops a Mathematics Program" *NAASP Bulletin* 52 (April 1968): 74–85.

16. James H. Zant, "Better Mathematics Teaching with Special Reference to Oklahoma," *Journal of Secondary Education* 39 (April 1964): 188–92.

17. NCTM, *Revolution in School Mathematics*, 17; minutes from Advisory Committee meeting of January 28, 1961, in "Drafts—Old Minutes," box 4.1/86–28/1 of SMSGR; H. Victor Crespy, "A Study of Curriculum Development in School Mathematics by National Groups, 1950–1966: Selected Programs" (Ph.D. diss., Temple University, 1969), 134.

18. "Some Present Problems of the School Mathematics Study Group," dated March 1965 in folder "Executive Committee 2 of 2," box 13.6/86–28/76, SMSGR.

19. Interview with Elsie Begle, box 4RM15, R. L. Moore Legacy Collection (hereafter Moore Collection), 1890–1900, 1920–2009, Archives of American Mathematics, Dolph Briscoe Center for American History, University of Texas at Austin, 13. Swain's text was published as Walter W. Hart, Veryl Schult, and Henry Swain, *First Year Algebra* (Boston: D. C. Heath, 1957).

20. Letter of July 4, 1963, in "Executive Committee 1961–1964," box 13.4/86–28/45, SMSGR.

21. Letter from E. G. Begle to Mary Dolciani, in folder "U-V 1959–1963, W 1962–1963," box 13.6/86–28/70, SMSGR. I am indebted to Trevor Crippen at Houghton Mifflin for basic contract information about the Modern Mathematics series.

22. E.g., an anonymous competitor said Houghton Mifflin was "walking away with the market"; as quoted in "New Math Book Sales Add Up: Publishers Find a Profitable Equation in Texts as Teaching Is Revolutionized," *Business Week*, April 10, 1965, 117–20, at 120; for 1970s figures, see Iris R. Weiss, *Report of the*

1977 National Survey of Science, Mathematics, and Social Studies Education SE 78–72 (Washington, DC: GPO, 1978), 90.

23. Letter from E. G. Begle to Austin MacCaffrey, from February 25, 1966, in folder "Advisory Board Minutes Exec. Committee 1 of 2," box 13.6/86–28/76, SMSGR. Despite Begle's pleas, the final newsletter listed books still available for purchase; none had been removed because no entirely suitable commercial texts had been found: SMSG, *Newsletter No. 24* (October 1966); *Newsletter No. 43* (August 1976), both in SMSGR.

24. Nearly all those who worked on writing elementary textbooks in summer sessions were affiliated with high schools or colleges; see "Advisory Board, Panels, and Writing Teams," box 13.5/86–28/64, SMSGR; Harry Schwartz, "The New Math Is Replacing Third 'R,'" *New York Times*, January 25, 1965, 18.

25. Paul C. Rosenbloom, "What Is Coming in Elementary Mathematics," *Educational Leadership* 18 (November 1960): 96–100, at 100; SMSG, *Newsletter No. 19* (September 1964); cf. Harry Schwartz, "'New Math' Leader Sees Peril in Haste," *New York Times*, December 31, 1964, 1 and 16.

26. Council for Basic Education, *Five Views of the "New Math"* (Washington, DC: CBE, 1965), 19; U.S. Department of Health, Education, and Welfare, Division of Educational Statistics and Bureau of Educational Research and Development, *Digest of Educational Statistics*, Bulletin 1963, no. 43 (Washington, DC: U.S. Government Printing Office, 1963), 26; Conference Board of the Mathematical Sciences National Advisory Committee on Mathematical Education, *Overview and Analysis of School Mathematics Grades K–12* (Washington, DC: CBMS, 1975), 77; while this latter survey of elementary teachers occurred after the spread of the new math textbooks, the authors note that teachers' average characteristics have held steady throughout the 1960s and 1970s; for certification, James Bryant Conant, *The Education of American Teachers*, new ed. (New York: McGraw-Hill, 1964 [1963]), 300–305; on 343 Conant cites Howard F. Fehr, "The Administration of Mathematics Education in the United States of America," *School Science and Mathematics* 55 (1955): 340–44.

27. For anxiety about these, and similar peculiarities of the elementary situation, see the folders on elementary mathematics, esp. "Elementary Centers Sept. 63–Feb. 64," box 13.4/86–28/45, SMSGR.

28. "Comments on First Grade Material" folder, box 13.4/86–28/48, SMSGR.

29. The various articles appear on pp. 1 and 18 of the *New York Times*, January 25, 1965; "New Math Book Sales Add Up: Publishers Find a Profitable Equation in Texts as Teaching Is Revolutionized," *Business Week* (April 10, 1965): 117–20; "New Math Your Children May Be Learning," *Good Housekeeping* 160 (January 1965): 132–33; Rhoda W. Bacmeister, "Preparing Preschoolers for the New Math: Just Playing with a Set of Blocks Gives Them Clues to What They'll Learn Later on in School," *Parents Magazine* 40 (September 1965): 64, 111–15; John L. Creswell, "Mom and Dad Study the New Math," *Parents Maga-*

zine 41 (September 1966): 69, 96–98; Nathan Cohn, "New Math at 33 1/3," *Saturday Review*, March 16, 1968, 55; Darrell Huff, "Understanding the New Math?" *Harper's Magazine* 231 (September 1965): 134–37.

30. See "SMSG Gross Sales Report—Yearly July 1/June 30," in folder "SMSG Financial Reports—Yearly July/June 30 1961–1970," box 13.6/86–28/68, SMSGR.

31. Joseph N. Payne, "The New Math and Its Aftermath, Grades K–8," in George Stanic and Jeremy Kilpatrick, eds., *A History of School Mathematics*, 1:559–98 (Reston, VA: National Council of Teachers of Mathematics, 2003), 569.

32. Robert W. Hayden, "A History of the 'New Math' Movement in the United States" (Ph.D. diss., Iowa State University, 1981), 179–87; J. David Lockard, ed., *Third Report of the Information Clearinghouse on New Science and Mathematics Curricula* (American Association for the Advancement of Science and the Science Teaching Center, University of Maryland, March 1965), esp. 35–67.

33. James Bryant Conant, *Education and Liberty: The Role of the Schools in a Modern Democracy* (Cambridge: Harvard University Press, 1953), 82–109; Frank B. Lindsay, "Mathematics Comes of Age in California High Schools," *California Journal of Secondary Education* 35 (February 1960): 79–83; F. Lindsay, "Recent Developments in High School Mathematics in California," *Journal of Secondary Education* 36 (December 1961): 478–85, esp. 481–82. Lindsay was the head of the Bureau of Secondary Education in the California Department of Education; both articles are from the same journal, but a title change had occurred in the interim. On "strands," see Janet Briggs Abbot, "Developing Mathematical Understanding by Implementing Strands Report," *California Education* 3 (November 1965): 9–11; "The 'New' Mathematics," *California Education* 1 (November 1963): 16, describes changes to the state's curriculum and cites the fall 1962 issue of the California Mathematics Council *Bulletin* on "strands" as the basic ideas of "modern" mathematics.

34. Hillel Black, *The American Schoolbook* (New York: Morrow, 1967), 34; John William Tebbel, *A History of Book Publishing in the United States*, 4 vols. (New York: R. R. Bowker, 1972–81), 4:493, 522.

35. Industry official as quoted in Black, *American Schoolbook*, on the unpaginated title page; for Scott, Foresman, see Black, *American Schoolbook*, 65–67.

36. M. Frank Redding, *Revolution in the Textbook Publishing Industry* (Washington, DC: National Education Association, 1963), esp. 11–14.

37. Adam Shapiro, "State Regulation of the Textbook Industry," in Adam R. Nelson and John L. Rudolph, eds., *Education and the Culture of Print in Modern America*, 173–90 (Madison: University of Wisconsin Press, 2010).

38. House Study Group of the Texas House of Representatives, "Texas Textbook Controversy," no. 101 (May 9, 1984), 2. This system, at least in California, was later changed in favor of more inclusive lists; see Raymond English, "The

Politics of Textbook Adoption," *Phi Delta Kappan* 62 (December 1980): 275–78, at 275–76. For Harper and Row's experience, see Black, *American Schoolbook*, 25–26.

39. B. G. Nunley, "Programed Instruction: Shift to 'New Math,'" *Texas Outlook* 47 (September 1963): 29; "New Math: A Story of Evolution," *Texas Outlook* 48 (April 1964): 33.

40. "New Math: A Story of Evolution"; for "leap of faith," Sam M. Gibbs, "New Math: Curriculum Changes," *Texas Outlook* 48 (April 1964): 34, 54; Jerry Irons, "Five Years after the Bomb: Or It's Time Again to Adopt New Elementary Math Books," *Texas Outlook* 54 (February 1970): 40.

41. While individual schools and the textbook adoption lists were clearly shifting by the early 1960s, some have dated the arrival of the new math in California to 1969 or 1970, when the state adopted the Houghton Mifflin series; see Frank Kendig, "Does the New Math Add Up?" *New York Times*, January 6, 1974, E14, 28, 32. Given the decentralized nature of education, even within states, it is hard to specify exactly when new math was introduced to a state. In any case, by the second half of the decade new textbooks were clearly widespread in California and Texas schools.

42. They remained so long after the 1960s: Cathy L. Seeley, "Mathematics Textbook Adoption in the United States," in George M. A. Stanic and Jeremy Kilpatrick, eds., *A History of School Mathematics*, 2:947–88 (Reston, VA: National Council of Teachers of Mathematics, 2003), esp. 972–73.

43. Thomas P. Ruff and Donald C. Orlich, "How Do Elementary-School Principals Learn about Curriculum Innovations?" *Elementary School Journal* 74 (April 1974): 389–92, at 390; Kenneth Easterday, "Some Wrong Slants Righted concerning 'New' Math," *Alabama School Journal* 82 (January 1965): 4–5; John F. Smith, "Case History, Charlotte-Mecklenburg," *NAASP Bulletin* 52 (April 1968): 65–73.

44. Such accounting is fraught with difficulties in the sense that numbers often depend upon how one counts—separating different versions or levels as individual books, etc. Nonetheless, even when one uses a conservative method such as unique author pairings, the available textbooks increased rapidly from the late 1950s through the middle 1960s. See the yearly *Textbooks in Print* published by R. R. Bowker Co.

45. For "foolhardy": Stephen S. Willoughby, "What Is the New Mathematics," *NAASP Bulletin* 52 (April 1968): 4–15, at 9; Richard P. Feynman, "New Textbooks for the New Mathematics," *California Education* 3 (1966): 8–14; for American Book Company: "New Math Book Sales Add Up," 120.

46. See 1962's *Textbooks in Print* (the 1962 publication listed the books published in the previous calendar year).

47. John Wagner, "The Objectives and Activities of the School Mathematics Study Group," *Mathematics Teacher* 53 (October 1960): 454–59, at 454; SMSG

[Max S. Bell et al.], *Studies in Mathematics*, vol. 9: *A Brief Course in Mathematics for Elementary School Teachers*, rev. ed. (Palo Alto: Stanford University Press, 1963), unnumbered first page.

48. E.g., "SMSG History" and "W. Wooton's SMSG History: Correspondence and Ms. Material," box 4.1/86–28/2, SMSGR, as well as the editing of John Mayor's letter to *Science* in "John R. Mayor," box 13.5/86–28/58, SMSGR.

49. "Curriculum Improvement in Mathematics," in folder "Basic Principles," box 4.1/86–28/5, SMSGR.

50. Speech in folder "Speeches—E. G. Begle," box 4.1/86–28/48, SMSGR.

51. E. G. Begle, "The School Mathematics Study Group," *NASSP Bulletin* 43 (May 1959): 26–31, at 27–28.

52. NCTM, *Revolution in School Mathematics*, 85.

53. Alvin E. Rolland, "Making the Switch to Modern Mathematics," *School and Community* 48 (May 1962): 29.

54. Arthur Beringause, "Search and Research in Education: Not Euclid Alone," *High Points* 44 (May 1962): 49–52, at 52.

55. Harry L. Phillips and Marguerite Kluttz, *Modern Mathematics and Your Child* (Washington, DC: U.S. Government Printing Office, 1965), esp. 1–3.

56. NCTM, "The Secondary Mathematics Curriculum: Report of the Secondary-School Committee of the National Council of Teachers of Mathematics," *Mathematics Teacher* 52 (May 1959): 389–417, at 390–92 (emphasis in original).

57. Donald F. Define, "Mathematics: Formulation of the Curriculum at Rich Township High School," *Clearing House* 36 (April 1962): 460–63, at 460.

58. B. R. Reardon, "I'm for the Modern Math. Here's Why," *Alabama School Journal* 83 (January 1966): 9.

59. For an examination of "modernization" as theory and ideology, see Nils Gilman, *Mandarins of the Future: Modernization Theory in Cold War America* (Baltimore: Johns Hopkins University Press, 2003); Michael E. Latham, *Modernization as Ideology: American Social Science and "National Building" in the Kennedy Era* (Chapel Hill: University of North Carolina Press, 2000); M. Latham, *The Right Kind of Revolution: Modernization, Development, and U.S. Foreign Policy from the Cold War to the Present* (Ithaca, NY: Cornell University Press, 2011), chaps. 1–2; an introductory overview of modernization's reach in the context of Cold War science generally is provided in Audra Wolfe, *Competing with the Soviets: Science, Technology, and the State in Cold War America* (Baltimore: Johns Hopkins University Press, 2013), chap. 4.

60. Daniel Lerner, with Lucille W. Pevsner, *The Passing of Traditional Society: Modernizing the Middle East* (Glencoe, IL: Free Press, 1958), esp. 104.

61. Paul C. Rosenbloom, ed., *Modern Viewpoints in the Curriculum: National Conference on Curriculum Experimentation* (New York: McGraw-Hill, 1964), esp. vii–ix.

62. Ibid., xi.

63. John F. Kennedy, "Message from the President of the United States Relative to an Educational Program" [February 9, 1962], reprinted ibid., 296–303.

64. U.S. Congress, Senate, Committee on Labor and Public Welfare, Subcommittee on Education, *Elementary and Secondary Education Act of 1965: Background Material with Related Presidential Recommendations*, 89th Cong., 1st sess., January 25, 1965, 12–14, 16.

65. As quoted in Tyack and Cuban, *Tinkering toward Utopia*, 5, which cites Henry Perkinson, *The Imperfect Panacea: American Faith in Education, 1865–1965* (New York: Random House, 1968); Hugh Davis Graham, *The Uncertain Triumph: Federal Education Policy in the Kennedy and Johnson Years* (Chapel Hill: University of North Carolina Press, 1984), 61–67; also Julie Roy Jeffrey, *Education for Children of the Poor: A Study of the Origins and Implementation of the Elementary and Secondary Education Act of 1965* (Columbus: Ohio State University Press, 1978), chaps. 2–3.

66. Kennedy, "Message from the President," 303.

67. NCTM, *Revolution in School Mathematics*, 10–11; interview with G. Baley Price, box 4RM19, Moore Collection, 41 (emphasis in original).

68. Doris Kearns [Goodwin], *Lyndon Johnson and the American Dream* (New York: Harper and Row, 1976), 210–11; Barbara Barksdale Clowse, *Brainpower for the Cold War: The Sputnik Crisis and National Defense Education Act of 1958* (Westport, CT: Greenwood Press, 1981), 158; Advisory Commission on Intergovernmental Relations, *Intergovernmentalizing the Classroom: Federal Involvement in Elementary and Secondary Education*, vol. 5: *The Federal Role in the Federal System: The Dynamics of Growth* (Washington, DC: Advisory Commission on Intergovernmental Relations, 1981), esp. 40–41.

69. Max Rafferty, *On Education* (New York: Devin-Adair, 1968), 77; Congress, Senate, Committee on Labor and Public Welfare, *Science and Education for National Defense*, 85th Cong., 2d sess., January 21, 23, 28–30; February 6, 7, 18–21, 24–27; March 3, 5, 6, 10–13, 1958, on p. 1229.

70. Congress, Senate, Committee on Labor and Public Welfare, *Science and Education for National Defense*, 85th Cong., 2nd sess., March 28, 1958, 1373.

71. On the rise of mathematical approaches in this period generally, see Paul Erickson et al., *How Reason Almost Lost Its Mind: The Strange Career of Cold War Rationality* (Chicago: University of Chicago Press, 2013); S. M. Amadae, *Rationalizing Capitalist Democracy: The Cold War Origins of Rational Choice Liberalism* (Chicago: University of Chicago Press, 2003); on the general role of mathematics in structuralism in this period, see David Aubin, "The Withering Immortality of Nicolas Bourbaki: A Cultural Connector at the Confluence of Mathematics, Structuralism, and the Oulipo in France," *Science in Context* 10 (1997): 297–342; and for a contemporary discussion, Jean Piaget, *Le structuralisme* (Paris: Presses universitaires de France, 1968). Of course, mathematics was not unique in this regard, even among curriculum programs, e.g., John Rudolph,

Scientists in the Classroom: The Cold War Reconstruction of Science Education (New York: Palgrave, 2002), esp. 99.

72. David Raymond Jardini, "Out of the Blue Yonder: The RAND Corporation's Diversification into Social Welfare Research, 1946–1968" (Ph.D. diss., Carnegie Mellon University, 1996), esp. 339–41; Alex Abella, *Soldiers of Reason: The RAND Corporation and the Rise of the American Empire* (Orlando, FL: Harcourt, 2008); Erickson et al., *How Reason Almost Lost Its Mind*; Jennifer S. Light, *From Warfare to Welfare: Defense Intellectual and Urban Problems in Cold War America* (Baltimore: Johns Hopkins University Press, 2003); Philip Mirowski, *Machine Dreams: Economics Becomes a Cyborg Science* (Cambridge: Cambridge University Press, 2002); Hunter Crowther-Heyck, *Herbert A. Simon: The Bounds of Reason in Modern America* (Baltimore: Johns Hopkins University Press, 2005); Paul N. Edwards, *The Closed World: Computers and the Politics of Discourse in Cold War America* (Cambridge: MIT Press, 1996); Theodore M. Porter, *Trust in Numbers: The Pursuit of Objectivity in Science and Public Life* (Princeton: Princeton University Press, 1995).

73. "The New Math: Does It Really Add Up?" *Newsweek*, May 10, 1965, 112–17, at 116.

74. "Inside Numbers," *Time*, January 31, 1964, 34–35, at 35.

Chapter 6. The Basic Subject

1. Francis J. Mueller, "Goals for School Mathematics? Educators and Parents Differ," *Virginia Journal of Education* 63 (January 1970): 21–22, at 21; Fehr, as quoted in Harry Schwartz, "The New Math Is Replacing Third 'R,'" *New York Times*, January 25, 1965, 18; Frank Kendig, "Does the New Math Add Up?" *New York Times*, January 6, 1974, E14, 28, 32; Helen Thomas, "Julie: Learning the New Math," *Washington Post*, January 14, 1971, G6.

2. Robert Davis, as quoted in Council for Basic Education, *Five Views of the "New Math"* (Washington, DC: CBE, 1965), 10.

3. Robert E. Reys, R. D. Kerr, and John W. Alspaugh, "Mathematics Curriculum Change in Missouri Secondary Schools," *School and Community* 56 (December 1969): 6–7, 9, at 6; David L. Angus and Jeffrey E. Mirel, "Mathematics Enrollments and the Development of the High School in the United States, 1910–1994," in George M. A. Stanic and Jeremy Kilpatrick, eds., *A History of School Mathematics*, 1:441–89 (Reston, VA: National Council of Teachers of Mathematics, 2003), esp. 467–71; Robert E. Stake and Jack A. Easley, Jr., *Case Studies in Science Education*, vol. 1: *The Case Reports* (Washington, DC: Government Printing Office, 1978).

4. There were many "guides" to the new math in this period for parents or other citizens confused about the intellectual and pedagogical basis of the re-

forms: e.g., Charles M. Baker, Helen Curran, and Mary Metcalf, *The "New" Math for Teachers and Parents of Elementary School Children* (San Francisco: Fearon Publishers, 1964); Irving Adler, *A New Look at Arithmetic* (New York: John Day, 1964). For examples of the sort of opportunities schools provided for parents, see the references in chap. 5, n. 29, above.

5. On second generation parents, see Thomas Fortune, "Schools Fighting to Regain Trust of Disenchanted Parents," *Los Angeles Times*, December 3, 1972, A1; and "Math Textbook Is Target of B.H. [Beverly Hills] Parents," *Los Angeles Times*, November 30, 1972, WS5; for Winnetka, Martha Elizabeth Hildebrandt, "Parental Attitudes about Mathematics Education: New Math in Winnetka" (Ph.D. diss., Northwestern University, 1976), esp. 83–106; and for "too far too fast," "New Math Book Sales Add Up: Publishers Find a Profitable Equation in Texts as Teaching Is Revolutionized," *Business Week*, April 10, 1965, 117–20, at 117.

6. Robert W. Hayden, "A History of the 'New Math' Movement in the United States" (Ph.D. diss., Iowa State University, 1981), 214; "Budget Proposals . . . Grant 18758," box 13.6/86–28/71, School Mathematics Study Group Records, 1958–77 (hereafter SMSGR), Archives of American Mathematics, Dolph Briscoe Center for American History, University of Texas at Austin.

7. The coverage of the closing of SMSG was generally positive, although acknowledging some controversy. See, for example, Fred M. Hechinger, "Time + Trial = Acceptance," *New York Times*, August 6, 1972, E7; "Lessons of the New Math," editorial, *Wall Street Journal*, August 17, 1972, 10.

8. Speech of John Ashbrook, Republican from Ohio, Congress, House, 93rd Cong., 2nd sess., *Congressional Record* 120, pt. 7 (March 27, 1974): 8523–26, at 8523–25; Mary C. Churchill, "A Drawback Is Found in 'New Math,'" *New York Times*, November 4, 1973, 129.

9. Jack McCurdy, "Second Look: The New Math—Compounding Old Problems," *Los Angeles Times*, September 27, 1973, 1, 28–29; J. McCurdy, "Back to 'Concrete' Addition, Subtraction: State Board Gives 'New Math' a 1–2 Punch," *Los Angeles Times*, May 10, 1974, A1; and "New Math Fails Simple Tasks," *Los Angeles Sentinel*, March 29, 1973, B8.

10. Chicago: Ruth Moss, "Perplexing Merry-Go-Round: Does the New Math Add Up to an 'A' or an 'F'?" *Chicago Tribune*, May 7, 1973, B13 (the first of a series of five articles); New York: New York State Commission on the Quality, Cost, and Financing of Elementary and Secondary Education, *The Fleischmann Report on the Quality, Cost, and Financing of Elementary and Secondary Education in New York State*, 3 vols. (New York: Published by the author, 1972); and Harold Faber, "Curriculum Revision: A Continuing Process," *New York Times*, January 8, 1973; New Hampshire: Kendig, "Does the New Math Add Up?" E14.

11. Morris Kline, *Why Johnny Can't Add: The Failure of the New Math* (New York: Vintage Books, 1973).

12. See, e.g., Edward Edelson, "Reforming the Numbers Racket," *Washington Post*, February 25, 1973, BW10; Harry Schwartz, "Dethroning the 'New Math,'" *New York Times*, July 20, 1973, 29; Normand H. Cole, "'New Math' under Attack," *Boston Globe*, August 19, 1973, 109; cf. the silent reception of J. Fang, *Numbers Racket: The Aftermath of the "New Math"* (Port Washington, NY: Kennikat Press, 1968).

13. These numbers and quotations are from Kendig, "Does the New Math Add Up?" but can be widely found in newspaper coverage of these years; for a standard account of the malaise, see Diane Ravitch, *The Troubled Crusade: American Education, 1945–1980* (New York: Basic Books, 1983), 311–12.

14. Jack McCurdy, "IQ Scores of California Pupils Drop for 6th Consecutive Year," *Los Angeles Times*, May 11, 1973, D1; Kendig, "Does the New Math Add Up?"

15. New York State Commission, *Fleischmann Report*, vol. 2, chap. 6.

16. Gina Bari Kolata, "Aftermath of the New Math: Its Originators Defend It," *Science*, n.s. 195, no. 4281 (March 4, 1977): 854–57.

17. Leonard J. Garigliano, "Arithmetic Computation Scores: Or Can Children in Modern Mathematics Program Really Compute?" *School Science and Mathematics* 75 (1975): 399–412.

18. Annegret Harnischfeger and David E. Wiley, "The Marrow of Achievement Test Score Declines," *Educational Technology* 16 (June 1976): 5–14; also Kolata, "Aftermath of the New Math."

19. Edward G. Begle and James W. Wilson, "Evaluation of Mathematics Programs," in National Society for the Study of Education, *Mathematics Education: Sixty-Sixth Yearbook of the National Society for the Study of Education*, 367–404 (Chicago: NSSE, 1970), 393–401; see also the Arlington, Va., superintendent as quoted in Jay Matthews, "New Math Baffles Old Mathematician," *Washington Post*, November 15, 1972, A1 and A13, at A13.

20. Thomas Fortune, "The New Math: Adherents Multiplying Rapidly as the Years Add Up," *Los Angeles Times*, October 24, 1971, OC1; McCurdy, "Second Look: The New Math," 28–29; Hechinger, "Time + Trial = Acceptance"; "The Price of Mathophobia," *Time*, March 17, 1967, 61; "How They're Teaching Math to Your Kids," *Changing Times* 23 (November 1969): 19–22, at 22.

21. Begle and Wilson, "Evaluation of Mathematics Programs"; the results of the study were published by SMSG in a series of pamphlets. See the listings in SMSG's *Newsletter*. Patrick Suppes, the director of Stanford's Institute for Mathematical Studies in the Social Sciences, once claimed that the computational ability of students was actually quite low in the 1930s and 1940s, so claims of declining scores in the 1960s were exaggerated, but I have not found the relevant evidence for this assertion; Patrick Suppes, "Adding Up the New Math," *PTA Magazine* 60 (October 1965): 8–10, at 8. On the lack of previous testing for comparisons between curricula, see Lee J. Cronbach, "Evaluation for Course

Improvement" [1963], in Robert Heath, ed., *New Curricula*, 231–48 (New York: Harper and Row, 1964); and Ellen Condliffe Lagemann, *An Elusive Science: The Troubling History of Education Research* (Chicago: University of Chicago Press, 2000), 192.

22. "NCTM Issues," box 13.6/86–28/76, SMSGR.

23. National Council of Teachers of Mathematics' Regional Orientation Conferences in Mathematics (NCTM), *The Revolution in School Mathematics: A Challenge for Administrators and Teachers* (Washington, DC: NCTM, 1961), 50.

24. McCurdy, "Second Look: The New Math," 29.

25. Seymour B. Sarason, *The Culture of the School and the Problem of Change* (Boston: Allyn and Bacon, 1971), 12–13, 115, and 181.

26. E.g., Alice Huettig and John M. Newell, "Attitudes toward Introduction of Modern Mathematics Program by Teachers with Large and Small Number of Years' Experience," *Arithmetic Teacher* 13 (February 1966): 125–30. For evidence from interviews, see Stake and Easley, *Case Studies in Science Education*, 1:17.29.

27. For enrollment numbers, see the U.S. Census: http://www.allcountries .org/uscensus/247_enrollment_in_public_and_private_schools.html; and the National Center for Education Statistics: http://nces.ed.gov/programs/digest/ d95/dtab038.asp. On Moynihan and Kerner, see Ellen Herman, *The Romance of American Psychology: Political Culture in the Age of Experts* (Berkeley: University of California Press, 1995), chaps. 7–8; James S. Coleman, *Equality of Educational Opportunity* (Washington, DC: Government Printing Office, 1966); on Jencks, Robert H. Haveman, *Poverty Policy and Poverty Research: The Great Society and the Social Sciences* (Madison: University of Wisconsin Press, 1987), 135–38; and "What the Schools Cannot Do," *Time*, April 16, 1973, 78–85.

28. Frank Newman, "The Era of Expertise: The Growth, The Spread and Ultimately the Decline of the National Commitment to the Concept of the Highly Trained Expert; 1945 to 1970" (Ph.D. diss., Stanford University, 1981), chaps. 6–7; Louis Menand, *The Marketplace of Ideas: Reform and Resistance in the American University* (New York: Norton, 2010), esp. 77; the phrase itself was, of course, a parody: David Halberstam, *The Best and the Brightest* (New York: Random House, 1972).

29. Jefferson Cowie, *Stayin' Alive: The 1970s and the Last Days of the Working Class* (New York: New Press, 2010), esp. 24.

30. McCurdy, "Back to 'Concrete' Addition, Subtraction."

31. Daniel P. Moynihan, *Maximum Feasible Misunderstanding: Community Action in the War on Poverty* (New York: Free Press, 1969), 163 and 203 (emphasis in original). Allen J. Matusow, *The Unraveling of America: A History of Liberalism in the 1960s* (New York: Harper and Row, 1984), esp. 225.

32. David C. King and Zachary Karabell, *The Generation of Trust: How the U.S. Military Has Regained the Public's Confidence since Vietnam* (Washing-

ton, DC: AEI Press, 2003), 2, as cited in James T. Patterson, *Restless Giant: The United States from Watergate to Bush v. Gore* (New York: Oxford University Press, 2005), chaps. 1–3, esp. 89.

33. Nils Gilman, *Mandarins of the Future: Modernization Theory in Cold War America* (Baltimore: Johns Hopkins University Press, 2003), chap. 7, esp. 244–49; Tom Engelhardt, *The End of Victory Culture: Cold War America and the Disillusioning of a Generation* (Amherst: University of Massachusetts Press, 1995).

34. Sharon Ghamari-Tabrizi, *The Worlds of Herman Kahn: The Intuitive Science of Thermonuclear War* (Cambridge: Harvard University Press, 2005); Rachel Carson, *Silent Spring* (Boston: Houghton Mifflin, 1962); a concise summary of these developments is found in Audra Wolfe, *Competing with the Soviets: Science, Technology, and the State in Cold War America* (Baltimore: Johns Hopkins University Press, 2013), chap. 7. See also Zuoyue Wang, *In Sputnik's Shadow: The President's Science Advisory Committee and Cold War America* (New Brunswick, NJ: Rutgers University Press, 2008); Brian Balogh, *Chain Reaction: Expert Debate and Public Participation in American Commercial Nuclear Power, 1945–1975* (New York: Cambridge University Press, 1991); and Daniel S. Greenberg, *The Politics of Pure Science* (New York: New American Library, 1967). There had been, of course, an undercurrent of distrust or unease with the scientist since at least the first use of a nuclear bomb. See, e.g., Paul S. Boyer, *By the Bomb's Early Light: American Thought and Culture at the Dawn of the Atomic Age* (New York: Pantheon, 1985), and Spencer R. Weart, *The Rise of Nuclear Fear* (Cambridge: Harvard University Press, 2012).

35. Daniel Sweeney, "Why Mohole Was No Hole," *American Heritage of Invention and Technology* 9 (1993): 54–63; Milton Lomask, *A Minor Miracle: An Informal History of the National Science Foundation* (Washington, DC: Government Printing Office, 1976), chap. 11; Willard Bascom, *A Hole in the Bottom of the Sea: The Story of the Mohole Project* (Garden City, NJ: Doubleday, 1961).

36. Erika Lorraine Milam, "Public Science of the Savage Mind: Contesting Cultural Anthropology in the Cold War Classroom," *Journal of the History of the Behavioral Sciences* 49 (2013): 306–30; Jamie Cohen-Cole, *The Open Mind: Cold War Politics and the Sciences of Human Nature* (Chicago: University of Chicago Press, 2014), chaps. 8 and 10; also, Lagemann, *An Elusive Science*, 172–76, and Peter B. Dow, *Schoolhouse Politics: Lessons from the Sputnik Era* (Cambridge: Harvard University Press, 1991). For Conlan's comments, see Congress, House, 94th Cong., 2nd sess., *Congressional Record* 122, pt. 3 (February 5, 1976): 2707–8, and pt. 7 (March 25, 1976): 7979.

37. Langton T. Crane, *The National Science Foundation and Pre-College Science Education: 1950–1975*, report prepared for the Committee on Science and Technology, House, 94th Cong., 2nd sess. (Washington, DC: Government Printing Office, 1975), 200–203.

38. Harry Schwartz, "The New Math Faces a Counter-Revolution," *New York Times*, January 8, 1973, 85.

39. Jerome S. Bruner, "*The Process of Education* Revisited," *Phi Delta Kappan* 53 (September 1971): 18–21.

40. Jonathan Kozol, *Death at an Early Age: The Destruction of the Hearts and Minds of Negro Children in the Boston Public Schools* (New York: Penguin, 1985 [1967]); Philip W. Jackson, *Life in Classrooms* (New York: Holt, Rinehart and Winston, 1968); Louis M. Smith and William Geoffrey, *The Complexities of an Urban Classroom: An Analysis toward a General Theory of Teaching* (New York: Holt, Rinehart and Winston, 1968).

41. It almost goes without mentioning that these authors emerged from distinct philosophical—and political—traditions and did not treat the school symmetrically. All nevertheless considered the school, like the workhouse, prison, and asylum, as a place where bodies and minds were disciplined. E.g., Erving Goffman, "On the Characteristics of Total Institutions" [1957], in E. Goffman, *Asylums: Essays on the Social Situation of Mental Patients and Other Inmates*, new ed., 1–124 (New Brunswick, NJ: Transaction, 2007 [1961]); Basil Bernstein, *Class, Codes, and Control*, vol. 3: *Towards a Theory of Educational Transmissions* (London: Routledge and Kegan Paul, 1975); Pierre Bourdieu and Jean-Claude Passeron, *Reproduction in Education, Society, and Culture*, trans. Richard Nice (London: Sage, 1977).

42. See, for example, the speech of John Ashbrook (n. 8) and the article of Edith Green, a former supporter of federal education legislation, in Congress, House, 93rd Cong., 2nd sess., *Congressional Record* 120, pt. 7 (March 27, 1974): 8523–26.

43. Ben Brodinsky, "Back to the Basics: The Movement and Its Meaning," *Phi Delta Kappan* 58 (March 1977): 522–27; Julie Roy Jeffrey, *Education for Children of the Poor: A Study of the Origins and Implementation of the Elementary and Secondary Education Act of 1965* (Columbus: Ohio State University Press, 1978), chap. 7.

44. E.g., Ralph W. Tyler, *The Florida Accountability Program: An Evaluation of Its Educational Soundness and Implementation* (Washington, DC: National Education Association, 1978).

45. Lagemann, *An Elusive Science*, 204–11; Lee Sproull, Stephen Weiner, and David Wolf, *Organizing an Anarchy: Belief, Bureaucracy, and Politics in the National Institute of Education* (Chicago: University of Chicago Press, 1978).

46. Crane, *National Science Foundation*, 5–8, 97–98, 101; Z. Wang, *In Sputnik's Shadow*, chaps. 15–16.

47. Crane, *National Science Foundation*, 9, 75–77, 98–105, 146–68.

48. Matthews, "New Math Baffles Old Mathematician," A1. In addition to being cited by numerous other media sources, the Shackelford episode was part of the arguments against the new math in Congress, e.g., New Hampshire senator

Norris Cotton's comments in Congress, Senate, 93rd Cong., 1st sess., *Congressional Record* 119, pt. 3 (February 1, 1973): 3076–77. For the election comparison, see R. W. Smith, "Letters to the Editor: Culturally It Adds Up," *Washington Post*, December 8, 1972, A27. The letters to the *Washington Post* editor were scattered over many days, but most were printed on November 24, 1972, A29.

49. "Nostalgia's Child: Back to the Basics," editor's page, *Phi Delta Kappan* 58 (March 1977): 521; George H. Gallup, "Ninth Annual Gallup Poll of the Public's Attitudes toward the Public Schools," *Phi Delta Kappan* 59 (September 1977): 33–36; for "most talked about," see Gene I. Maeroff, "Issue and Debate: The Return to Fundamentals in the Nation's Schools," *New York Times*, December 6, 1975, 58; "Position Statements on Basic Skills," *Mathematics Teacher* 71 (February 1978): 147–55.

50. Council for Basic Education, *Five Views of the "New Math"*; National School Boards Association Research Report, *Back to Basics*, Report 1978–1 (Washington, DC: NSBA, 1978), esp. 8.

51. Fortune, "Schools Fighting to Regain Trust," A1. Many similar examples exist, e.g., Meg O'Connor, "At '3-Rs' School on Southside, Goal Strictly Education," *Chicago Tribune*, September 19, 1976, 45; Edward B. Fiske, "Suburban Schools Are Evolving 'Basic' Curriculums Geared to 1970's," *New York Times*, June 15, 1977, 45.

52. Lynn Lilliston, "Getting Back to Schooling Fundamentals," *Los Angeles Times*, January 30, 1974, E1 and 4; "Back to Basics in the Schools," *Newsweek*, October 21, 1974, 87–95; perhaps tellingly, when the article was reprinted in *Reader's Digest*, the section on the "progressive" school was simply omitted: "Back to Basics in the Schoolhouse," *Reader's Digest* 106 (February 1975): 149–52.

53. E.g., Lorraine Daston, "Enlightenment Calculations," *Critical Inquiry* 21 (1994): 182–202; Matthew L. Jones, *The Good Life in the Scientific Revolution: Descartes, Pascal, Leibniz, and the Cultivation of Virtue* (Chicago: University of Chicago Press, 2006).

54. Stake and Easley, *Case Studies in Science Education* , 1:1.29, 2.11, 1.104–7; Robert E. Stake and Jack A. Easley, Jr., *Case Studies in Science Education*, vol. 2: *Design, Overview and General Findings* (Washington, DC: U.S. Government Printing Office, 1978), chap. 12, pp. 33–34. Researchers eventually published seven volumes of data, with an overview in Stanley L. Helgeson, Patricia E. Blosser, and Robert W. Howe, *The Status of Pre-College Science Mathematics, and Social Studies Practices in U.S. Schools: An Overview and Summaries of Three Studies*, SE 78–71 (Washington, DC: U.S. Government Printing Office, 1978).

55. Stake and Easley, *Case Studies in Education*, 1:1.31, 33, 107. Although the view that few students could handle the new math was nearly universal, some teachers preferred the new math as a curriculum, particularly in high schools; see, e.g., ibid., 1:4.29.

56. For "lumping," see, e.g., B. D. Colen, "Conservative School Set in Pr. George's," *Washington Post*, December 3, 1974, A1 and A5, at A5; Fred M. Hechinger, "Where Have All the Innovations Gone?" *New York Times*, November 16, 1975, ED30.

57. "Back to Basics in the Schools," 95.

58. Edelson, "Reforming the Numbers Racket," BW10.

59. Burton Yale Pines, *Back to Basics: The Traditionalist Movement That Is Sweeping Grass-Roots America* (New York: William Morrow, 1982), esp. 99–129.

60. Lisa McGirr, *Suburban Warriors: The Origins of the New American Right* (Princeton: Princeton University Press, 2001), esp. 70–75, 179–81; Michelle M. Nickerson, *Mothers of Conservatism: Women and the Postwar Right* (Princeton: Princeton University Press, 2012), chap. 3; Sara Diamond, *Not by Politics Alone: The Enduring Influence of the Christian Right* (New York: Guilford Press, 1998).

61. J. Dan Marshall, "With a Little Help from Some Friends: Publishers, Protestors, and Texas Textbook Decisions," in Michael W. Apple and Linda K. Christian-Smith, eds., *The Politics of the Textbook*, 56–77 (New York: Routledge, 1991); James C. Hefley, *Are Textbooks Harming Your Children? Norma and Mel Gabler Take Action and Show You How* (Milford, MI: Mott Media, 1979), esp. 218; the Gablers' influence extended well beyond their particular crusade, e.g., Herbert Ira London, *Why Are They Lying to Our Children?* (New York: Stein and Day, 1984).

62. Parents' resentment of experts is discussed in more detail in Jonathan Zimmerman, *Whose America? Culture Wars in the Public Schools* (Cambridge: Harvard University Press, 2002), 200–206; see also William Martin, *With God on Our Side: The Rise of the Religious Right in America* (New York: Broadway, 1996), chaps. 5–6. The general literature on textbook controversies is large. See, e.g., Joseph Moreau, *Schoolbook Nation: Conflicts over American History Textbooks from the Civil War to the Present* (Ann Arbor: University of Michigan Press, 2003); Diane Ravitch, *The Language Police: How Pressure Groups Restrict What Students Learn* (New York: Alfred A. Knopf, 2003); Paul S. Boyer, *Purity in Print: Book Censorship in America from the Gilded Age to the Computer Age*, 2nd ed. (Madison: University of Wisconsin Press, 2002); Robert Lerner, Althea K. Nagai, and Stanley Rothman, *Molding the Good Citizen: The Politics of High School History Texts* (Westport, CT: Praeger, 1995); Joan Del-Fattore, *What Johnny Shouldn't Read: Textbook Censorship in America* (New Haven: Yale University Press, 1992); James Moffett, *Storm in the Mountains: A Case Study of Censorship, Conflict, and Consciousness* (Carbondale: Southern Illinois University Press, 1988); Frances Fitzgerald, "A Reporter at Large: A Disagreement in Baileyville," *New Yorker*, January 16, 1984, 47–90; Frances Fitzgerald, *America Revised: History Schoolbooks in the Twentieth Century* (Boston:

Little, Brown, 1979); Edward B. Jenkinson, *Censors in the Classroom: The Mind Benders* (Carbondale: Southern Illinois University Press, 1979); Dorothy Nelkin, *Science Textbook Controversies and the Politics of Equal Time* (Cambridge: MIT Press, 1977).

63. Diamond, *Not by Politics Alone*, 65.

64. Robert C. Snider, "Back to the Basics?" NEA Position Paper (August 1978), 6–9; "Nostalgia's Child," 521; Fred M. Hechinger, "The Back-to-the-Basics Impact," *Today's Education* 67 (February–March 1978): 31–32. Even Ford Motors introduced an advertising slogan for the 1970s which emphasized the cultural pervasiveness of such sentiments: "When you get back to basics, you get back to Ford" (*New York Times*, November 30, 1971, 41).

65. McGirr, *Suburban Warriors*, 84; Cf. Jonathan Reider, "The Rise of the 'Silent Majority,'" in Steve Fraser and Gary Gerstle, eds., *The Rise and Fall of the New Deal Order, 1930–1980*, 243–68 (Princeton: Princeton University Press, 1989).

66. For the expanding literature on the rise of the "right," see Kim Phillips-Fein, "Conservatism: A State of the Field," *Journal of American History* 98 (2011): 723–43, and Julian E. Zelizer, "Rethinking the History of American Conservatism," *Reviews in American History* 38 (2010): 367–92.

67. "Decentralization" itself became an object of federal study: Robert K. Yin and Douglas Yates, *Street-Level Governments: Assessing Decentralization and Urban Services (An Evaluation of Policy Research)* (Santa Monica: Rand, 1974), as discussed in Jennifer S. Light, *From Warfare to Welfare: Defense Intellectuals and Urban Problems in Cold War America* (Baltimore: Johns Hopkins University Press, 2003), chap. 6. For the phrase "decade of the neighborhood," see Suleiman Osman, "The Decade of the Neighborhood," in Bruce J. Schulman and Julian E. Zelizer, eds., *Rightward Bound: Making America Conservative in the 1970s*, 106–27 (Cambridge: Harvard University Press, 2008), 106–27; S. Osman, *The Invention of Brownstone Brooklyn: Gentrification and the Search for Authenticity in Postwar New York* (New York: Oxford University Press, 2011). For divisions among conservatives, see Brodinsky, "Back to the Basics."

68. For Boston, see Ronald P. Formisano, *Boston against Busing: Race, Class, and Ethnicity in the 1960s and 1970s* (Chapel Hill: University of North Carolina Press, 1991); and Adam R. Nelson, *The Elusive Ideal: Equal Educational Opportunity and the Federal Role in Boston's Public Schools, 1950–1985* (Chicago: University of Chicago Press, 2005); for Orange County's "Not In My Back Yard" protests, see McGirr, *Suburban Warriors*, esp. chap. 6.

69. On the emergence of "liberal consensus," see Godfrey Hodgson, *America in Our Time* (Garden City, NJ: Doubleday, 1976); the degree of actual consensus was a matter of debate, particularly on divisive issues like race: Gary Gerstle, "Race and the Myth of the Liberal Consensus," *Journal of American History* 82 (1995): 579–86.

70. Larry Green, "'New Math' in New Guise: The Old Math," *Los Angeles Times*, November 28, 1977, A6 and 18, at A6.

71. Iver Peterson, "Nation's Schools Renewing Stress on the Basics," *New York Times*, March 3, 1975, 1 and 38, at 38.

72. Stake and Easley, *Case Studies in Education*, 2:12.34–35, 13.25.

73. On the fate of "structural thinking" generally, see Daniel T. Rodgers, *Age of Fracture* (Cambridge: Harvard University Press, 2011).

74. The Gablers' quotation is from Martin, *With God on Our Side*, 121, who cites Moffett, *Storm in the Mountains*.

Epilogue

1. For the "traditional in nature" algebra text, see D. Franklin Wright and Bill D. New, *Introductory Algebra* (Boston: Allyn and Bacon, 1981), xi and chap. 1; similar post-new-math examples include, among many others, Daniel L. Auvil and Charles Poluga, *Elementary Algebra* (Reading, MA: Addison-Wesley, 1978); Richard Johnson, Margaret Ullmann Miller, and Steven D. Kerr, *Elementary Algebra* (Menlo Park, CA: Benjamin/Cummings, 1981); and Daniel D. Benice, *Arithmetic and Algebra*, 2nd ed. (Englewood Cliffs, NJ: Prentice-Hall, 1979). For a range of publishers' approaches prior to the curricular reforms, see, e.g., Daniel W. Snader, *Algebra: Meaning and Mastery, Book 1* (Philadelphia: John C. Winston Company, 1949); Raleigh Schorling, Rolland R. Smith, and John R. Clark, *Algebra: First Course* (New York: World Book Company, 1949); William O. Shute et al., *Elementary Algebra* (New York: American Book Company, 1956); Walter W. Hart, Veryl Schult, and Henry Swain, *First Year Algebra* (Boston: D. C. Heath, 1957); and Frank M. Morgan and Burnham L. Paige, *Algebra 1* (New York: Henry Holt, 1958). There were, of course, a handful of "modern" texts prior to SMSG as well, most notably one by a later reformer, Bruce E. Meserve, *Fundamental Concepts of Algebra* (Cambridge, MA: Addison-Wesley, 1953).

2. Gina Bari Kolata, "Aftermath of the New Math: Its Originators Defend It," *Science*, n.s. 195, no. 4281 (March 4, 1977): 854–57; Ross Taylor, "NACOME: Implications for Teaching K–12," *Mathematics Teacher* 69 (October 1976): 458–63; David L. Elliott, "Textbooks and the Curriculum in the Postwar Era: 1950–1980," in David L. Elliott and Arthur Woodward, eds., *Textbooks and Schooling in the United States: Eighty-ninth Yearbook of the National Society for the Study of Education*, 42–55 (Chicago: University of Chicago Press, 1990).

3. On the "math wars," especially in connection with the nearly simultaneous "reading wars," see Tom Loveless, ed., *The Great Curriculum Debate: How Should We Teach Reading and Math?* (Washington, DC: Brookings Institution Press, 2001), esp. chapters by Gail Burrill, Michael Battista, Richard Askey, and Tom Loveless; also Douglas B. McLeod, "From Consensus to Controversy: The

Story of the NCTM *Standards*," in George M. A. Stanic and Jeremy Kilpatrick, eds., *A History of School Mathematics*, 1:753–818 (Reston, VA: NCTM, 2003), at 798–800.

4. For an early example of this cyclic nature (and for the contextualization of the continued test score decline), see "Why Johnny Can't Add," *Newsweek*, September 24, 1979, 72. The calculator debate was afoot even before the new math controversy had died down: NCTM recommended in 1976 that computers be available to students and in 1980 that calculators be available: James T. Fey and Anna O. Graeber, "From the New Math to the *Agenda for Action*," in Stanic and Kilpatrick, *A History of School Mathematics*, 1:521–58, at 547–49.

5. Fey and Graeber, "From the New Math to the *Agenda for Action*"; see also Arthur Coxford, "Mathematics Curriculum Reform: A Personal View," and Zalman Usiskin, "A Personal History of the UCSMP Secondary School Curriculum, 1960–1999," in Stanic and Kilpatric, *History of School Mathematics*, 1:599–621 and 1:673–736, respectively.

6. National Council of Teachers of Mathematics, *Commission on Standards for School Mathematics* (Reston, VA: NCTM, 1989), esp. 2; Max S. Bell, "What Does 'Everyman' Really Need from School Mathematics?" *Mathematics Teacher* 67 (March 1974): 196–203. Bell later joined the University of Chicago School Mathematics Project.

7. NCTM, *Commission on Standards*, 1–3. On explicit connections between the new math and the *Standards*, see Angela L. E. Walmsley, "A History of the 'New Mathematics' Movement and Its Relationship with the National Council of Teachers of Mathematics Standards" (Ph.D. diss., Saint Louis University, 2001).

8. National Commission on Excellence in Education, *A Nation at Risk: The Imperative for Educational Reform* (Washington, DC: U.S. Government Printing Office, 1983). McLeod, "From Consensus to Controversy," 771.

9. Ibid., 796–97; Douglas B. McLeod et al., "Setting the Standards: NCTM's Role in the Reform of Mathematics Education," in Senta A. Raizen and Edward D. Britton, eds., *Case Studies of U.S. Innovations in Mathematics Education*, 3:13–132 (Dordrecht: Kluwer, 1996), 81; McLeod, "From Consensus to Controversy"; Diane Ravitch, *National Standards in American Education: A Citizen's Guide* (Washington, DC: Brookings Institution Press, 1995).

10. See, e.g., the statements of "Mathematically Correct," which were collected on the website mathematicallycorrect.com until it went offline around April 2013.

11. McLeod, "From Consensus to Controversy," 776–77, and McLeod et al., "Setting the Standards." More generally on the complicated politics of curriculum reform in this period, see Suzanne M. Wilson, *California Dreaming: Reforming Mathematics Education* (New Haven: Yale University Press, 2003).

Bibliography

Main Archival Sources

American Mathematical Society Archives. Brown University, Providence, RI.

Newton Public School Archives. Newton, MA.

R. L. Moore Legacy Collection (Moore Collection). 1890–1900, 1920–2009. Archives of American Mathematics, Dolph Briscoe Center for American History, University of Texas at Austin.

Roy E. Larsen Collection. Special Collections, Monroe C. Gutman Library, Harvard Graduate School of Education.

School Mathematics Study Group Records (SMSGR). 1958–77. Archives of American Mathematics, Dolph Briscoe Center for American History, University of Texas at Austin.

Published Sources

Abbott, Janet Briggs. "Developing Mathematical Understanding by Implementing Strands Report." *California Education* 3 (November 1965): 9–11.

Abella, Alex. *Soldiers of Reason: The RAND Corporation and the Rise of the American Empire*. Orlando, FL: Harcourt, 2008.

Adam, Alison. *Artificial Knowing: Gender and the Thinking Machine*. London: Routledge, 1998.

Adler, Irving. *A New Look at Arithmetic*. New York: John Day, 1964.

Adorno, T. W., Else Frenkel-Brunswik, Daniel J. Levinson, and R. Nevitt Sanford. *The Authoritarian Personality*. New York: Harper, 1950.

Advisory Commission on Intergovernmental Relations. *Intergovernmentalizing the Classroom: Federal Involvement in Elementary and Secondary Education*, vol. 5: *The Federal Role in the Federal System: The Dynamics of*

Growth. Washington, DC: Advisory Commission on Intergovernmental Relations, 1981.

Albert, A. Adrian. *Structure of Algebras.* New York: American Mathematical Society, 1939.

Alexanderson, G. L. "An Interview with Morris Kline: Part 2." *Two-Year College Mathematics Journal* 10 (1979): 259–64.

Amadae, S. M. *Rationalizing Capitalist Democracy: The Cold War Origins of Rational Choice Liberalism.* Chicago: University of Chicago Press, 2003.

Angus, David L., and Jeffrey E. Mirel. "Mathematics Enrollments and the Development of the High School in the United States, 1910–1994." In George M. A. Stanic and Jeremy Kilpatrick, eds., *A History of School Mathematics,* 1:441–89. Reston, VA: National Council of Teachers of Mathematics, 2003.

Aquinas, Thomas. *Summa Theologiae.* 61 vols. Cambridge: Cambridge University Press, 2006.

Ascham, Roger. *The Scholemaster.* London, 1579.

Aubin, David. "The Withering Immortality of Nicholas Bourbaki: A Cultural Connector at the Confluence of Mathematics, Structuralism, and the Oulipo in France." *Science in Context* 10 (1997): 297–342.

Auvil, Daniel L., and Charles Poluga. *Elementary Algebra.* Reading, MA: Addison-Wesley, 1978.

"Back to Basics in the Schoolhouse." *Reader's Digest* 106 (February 1975): 149–52.

"Back to Basics in the Schools." *Newsweek,* October 21, 1974, 87–95.

Bacmeister, Rhoda W. "Preparing Preschoolers for the New Math: Just Playing with a Set of Blocks Gives Them Clues to What They'll Learn Later on in School." *Parents Magazine* 40 (September 1965): 64, 111–15.

Bailyn, Bernard. *Education in the Formation of American Society.* Chapel Hill: University of North Carolina Press, 1960.

Baker, Charles M., Helen Curran, and Mary Metcalf. *The "New" Math for Teachers and Parents of Elementary School Children.* San Francisco: Fearon Publishers, 1964.

Balogh, Brian. *Chain Reaction: Expert Debate and Public Participation in American Commercial Nuclear Power, 1945–1975.* New York: Cambridge University Press, 1991.

Bascom, Willard. *A Hole in the Bottom of the Sea: The Story of the Mohole Project.* Garden City, NJ: Doubleday, 1961.

Beam, Alex. *A Great Idea at the Time: The Rise, Fall, and Curious Afterlife of the Great Books.* New York: Public Affairs, 2008.

Beaulieu, Liliane. "Bourbaki's Art of Memory." *Osiris,* 2nd ser., 14 (2002): 219–51.

Beberman, M. *An Emerging Program of Secondary School Mathematics.* Cambridge: Harvard University Press, 1958.

Begle, E. [Edward] G. *Critical Variables in Mathematics Education: Findings from a Survey of the Empirical Literature.* Washington, DC: MAA and NCTM, 1979.

——. "Remarks on the Memorandum 'On the Mathematics Curriculum of the High School.'" *American Mathematical Monthly* 69 (May 1962): 425–26.

——. "Review of *Why Johnny Can't Add.*" *National Elementary Principal* 53 (January–February 1974): 26–31.

——. "The School Mathematics Study Group." *NASSP Bulletin* 43 (May 1959): 26–31.

——. "The School Mathematics Study Group." *Mathematics Teacher* 51 (December 1958): 618.

——. "Some Lessons Learned by S.M.S.G." *Mathematics Teacher* 66 (1973): 207–14.

Begle, Edward G., and James W. Wilson. "Evaluation of Mathematics Programs." In National Society for the Study of Education, *Mathematics Education: Sixty-Sixth Yearbook of the National Society for the Study of Education*, 367–404. Chicago: NSSE, 1970.

Bell, Eric Temple. *Mathematics: Queen and Servant of Science.* New York: McGraw-Hill, 1951.

Bell, James A. "Trends in Secondary Mathematics in Relation to Psychological Theories: 1893–1970." Ph.D. diss., University of Oklahoma, 1971.

Bell, Max S. "What Does 'Everyman' Really Need from School Mathematics?" *Mathematics Teacher* 67 (March 1974): 196–203.

Benice, Daniel D. *Arithmetic and Algebra.* 2nd ed. Englewood Cliffs, NJ: Prentice-Hall, 1979.

Bensaude-Vincent, Bernadette. "Textbooks on the Map of Science Studies." *Science and Education* 15 (2006): 667–70.

Beringause, Arthur. "Search and Research in Education: Not Euclid Alone." *High Points* 44 (May 1962): 49–52.

Bernstein, Basil B. *Class, Codes, and Control.* 3 vols. London: Routledge and Kegan Paul, 1971–75.

Bestor, Arthur. *Educational Wastelands: The Retreat from Learning in Our Public Schools.* 2nd ed. Urbana: University of Illinois Press, 1985 [1953].

Bird, Kai, and Martin J. Sherwin. *American Prometheus: The Triumph and Tragedy of J. Robert Oppenheimer.* New York: Vintage, 2005.

Birkhoff, George D. "A Set of Postulates for Plane Geometry, Based on Scale and Protractor." *Annals of Mathematics*, 2nd ser., 33 (1932): 329–45.

Birkhoff, George David, and Ralph Beatley. *Geometry.* Prelim. ed. Boston: Spaulding-Moss, 1933.

Black, Hillel. *The American Schoolbook.* New York: Morrow, 1967.

Blass, Thomas. *The Man Who Shocked the World: The Life and Legacy of Stanley Milgram.* New York: Basic Books, 2004.

Bloor, David. "Wittgenstein and Mannheim on the Sociology of Mathematics." *Studies in History and Philosophy of Science* 4 (1973): 173–91.

———. *Wittgenstein, Rules and Institutions.* London: Routledge, 1997.

Boaler, Jo. "The Stereotypes about Math That Hold Americans Back." *Atlantic*, November 12, 2013; http://www.the atlantic.com/education/archive/2013/11/.

Bourbaki, Nic[h]olas. "L'architecture des mathématiques." In F. Le Lionnais, ed., *Les grands courants de la pensée mathématique*, 35–47. [Marseille]: Cahiers du Sud, 1948.

———. "The Architecture of Mathematics." *American Mathematical Monthly* 57 (1950): 221–32.

Bourdieu, Pierre. "Systems of Education and Systems of Thought." In Michael F. D. Young, ed., *Knowledge and Control: New Directions for the Sociology of Education*, 189–207. London: Collier Macmillan, 1971.

Bourdieu, Pierre, and Jean-Claude Passeron. *Reproduction in Education, Society, and Culture.* Trans. Richard Nice. London: Sage, 1977.

Boyer, Paul S. *By the Bomb's Early Light: American Thought and Culture at the Dawn of the Atomic Age.* New York: Pantheon, 1985.

———. *Purity in Print: Book Censorship in America from the Gilded Age to the Computer Age.* 2nd ed. Madison: University of Wisconsin Press, 2002.

Brainerd, Charles J. *Piaget's Theory of Intelligence.* Englewood Cliffs, NJ: Prentice-Hall, 1978.

Brodinsky, Ben. "Back to the Basics: The Movement and Its Meaning." *Phi Delta Kappan* 58 (March 1977): 522–27.

Browder, Felix E. "The Stone Age of Mathematics on the Midway." In Peter Duren, ed., *A Century of Mathematics in America*, 2:191–93. Providence: American Mathematical Society, 1989.

Brownell, William. "Principles in Curriculum Construction." In SMSG, *Report of a Conference on Elementary School Mathematics*, 15–17. New Haven: Yale University Press, 1959.

Brownell, William A., and Harold E. Moser. *Meaningful vs. Mechanical Learning: A Study in Grade III Subtraction.* Durham, NC: Duke University Press, 1949.

Bruner, Jerome. *The Process of Education.* New York: Vintage Books 1963 [1960].

———. "*The Process of Education* Revisited." *Phi Delta Kappan* 53 (September 1971): 18–21.

Bush, Vannevar. *Science: The Endless Frontier.* Washington, DC: Government Printing Office, 1945.

Cajori, Florian. *A History of Mathematical Notations.* 2 vols. Chicago: Open Court, 1928.

Callan, Mary Ann. "Crisis Ahead in Education." *Los Angeles Times*, October 22, 1957, A1.

Cambridge Conference on School Mathematics. *Goals for School Mathematics.* Boston: Houghton Mifflin, 1963.

Capshew, James H. *Psychologists on the March: Science, Practice, and Professional Identity in America, 1929–1969.* Cambridge: Cambridge University Press, 1999.

Carrier, George F., R. Courant, Paul Rosenbloom, and C. N. Yang. "Applied Mathematics: What Is Needed in Research and Education: A Symposium." *SIAM Review* 4 (1962): 297–320.

Carruthers, Susan L. *Cold War Captives: Imprisonment, Escape, and Brainwashing.* Berkeley: University of California Press, 2009.

Carson, John. *The Measure of Merit: Talents, Intelligence, and Inequality in the French and American Republics, 1750–1940.* Princeton: Princeton University Press, 2007.

Carson, Rachel. *Silent Spring.* Boston: Houghton Mifflin, 1962.

Cartan, Henri. "Nicholas Bourbaki and Contemporary Mathematics" [1958]. *Mathematical Intelligencer* 2 (1980): 175–87.

Cartan, Henri, Gustave Choquet, and Jacques Dixmier. *Structures algébriques et structures topologiques.* Geneva: Enseignement mathématique, 1958.

Chall, Jeanne S. *Learning to Read: The Great Debate.* 3rd ed. Fort Worth: Harcourt Brace College Publishers, 1996.

Churchill, Mary C. "A Drawback Is Found in 'New Math.'" *New York Times,* November 4, 1973, 129.

Clarke, Sabine. "Pure Science with a Practical Aim: The Meanings of Fundamental Research in Britain, circa 1916–1950." *Isis* 101 (2010): 285–311.

Clowse, Barbara Barksdale. *Brainpower for the Cold War: The Sputnik Crisis and National Defense Education Act of 1958.* Westport, CT: Greenwood Press, 1981.

Cohen, Lizabeth. *A Consumers' Republic: The Politics of Consumption in Postwar America.* New York: Vintage Books, 2004.

Cohen, Patricia Cline. *A Calculating People: The Spread of Numeracy in Early America.* Chicago: University of Chicago Press, 1982.

———. "Numeracy in Nineteenth-Century America." In George M. A. Stanic and Jeremy Kilpatrick, eds., *A History of School Mathematics,* 1:43–76. Reston, VA: National Council of Teachers of Mathematics, 2003.

Cohen-Cole, Jamie. "The Creative American: Cold War Salons, Social Science, and the Cure for Modern Society." *Isis* 100 (2009): 219–62.

———. *The Open Mind: Cold War Politics and the Sciences of Human Nature.* Chicago: University of Chicago Press, 2014.

———. "The Reflexivity of Cognitive Science: The Scientist as Model of Human Nature." *History of the Human Sciences* 18 (2005): 107–39.

———. "Thinking about Thinking in Cold War America." Ph.D. diss., Princeton University, 2003.

Cohn, Nathan. "New Math at 33 1/3." *Saturday Review,* March 16, 1968, 55.

Colburn, Warren. *First Lessons in Arithmetic, on the Method of Pestalozzi, with Some Improvements.* Boston: Cummings, Hilliard, 1825.

"Cold War of the Classrooms." *Science Digest* 39 (February 1956): 33.

Cole, Normand H. "'New Math' under Attack." *Boston Globe,* August 19, 1973, 109.

Coleman, James S. *Equality of Educational Opportunity.* Washington, DC: Government Printing Office, 1966.

Colen, B. D. "Conservative School Set in Pr. George's." *Washington Post,* December 3, 1974, A1, A5.

College Entrance Examination Board. *Report of the Commission on Mathematics: Program for College Preparatory Mathematics.* New York: College Entrance Education Board, 1959.

Collins, H. M. *Changing Order: Replication and Induction in Scientific Practice.* 2nd ed. Chicago: University of Chicago Press, 1992.

Commission on the Secondary School Curriculum of the Progressive Education Association. *Mathematics in General Education: A Report of the Committee on the Function of Mathematics in General Education.* New York: D. Appleton-Century, 1940.

Committee on Support of Research in the Mathematical Sciences. *The Mathematical Sciences: A Report.* Washington, DC: National Academy of Sciences, 1968.

Committee on the Objectives of a General Education in a Free Society. *General Education in a Free Society: Report of the Harvard Committee.* Cambridge: Harvard University Press, 1945.

Conant, James B. *The American High School Today: A First Report to Interested Citizens.* New York: McGraw-Hill, 1959.

———. *The Child, the Parent, and the State.* Cambridge: Harvard University Press, 1959.

———. *Education and Liberty: The Role of the Schools in a Modern Democracy.* Cambridge: Harvard University Press, 1953.

———. *The Education of American Teachers.* 2nd ed. New York: McGraw-Hill, 1964 [1963].

Conference Board of the Mathematical Sciences National Advisory Committee on Mathematical Education. *Overview and Analysis of School Mathematics Grades K–12.* Washington, DC: CBMS, 1975.

Corry, Leo. *David Hilbert and the Axiomatization of Physics (1898–1918): From "Grundlagen der Geometrie" to "Grundlagen der Physik."* Dordrecht: Kluwer Academic, 2004.

———. *Modern Algebra and the Rise of Mathematical Structures.* Basel: BirkhäuserVerlag, 1996.

Council for Basic Education. *Five Views of the "New Math."* Washington, DC: CBE, 1965.

Courant, Richard, and Herbert Robbins. *What Is Mathematics?* New York: Oxford University Press, 1941.

Cowie, Jefferson. *Stayin' Alive: The 1970s and the Last Days of the Working Class.* New York: New Press, 2010.

Coxford, Arthur. "Mathematics Curriculum Reform: A Personal View." In George M. A. Stanic and Jeremy Kilpatrick, eds., *A History of School Mathematics,* 1:599–621. Reston, VA: National Council of Teachers of Mathematics, 2003.

Crane, Langton T. *The National Science Foundation and Pre-College Science Education: 1950–1975.* Report prepared for the Committee on Science and Technology, House, 94th Cong., 2nd sess. Washington, DC: Government Printing Office, 1975.

Cremin, Lawrence A. *The American Common School: An Historic Conception.* New York: Teachers College, 1951.

———. *The Genius of American Education.* New York: Vintage Books, 1965.

———. *The Transformation of the School: Progressivism in American Education, 1876–1957.* New York: Alfred A. Knopf, 1969.

Crespy, H. Victor. "A Study of Curriculum Development in School Mathematics by National Groups, 1950–1966: Selected Programs." Ph.D. diss., Temple University, 1969.

Creswell, John L. "Mom and Dad Study the New Math." *Parents Magazine* 41 (September 1966): 69, 96–98.

"Crisis in Education: Schoolboys Point up a U.S. Weakness." *Life,* March 24, 1958, 25–35.

Cronbach, Lee J. "Evaluation for Course Improvement" [1963]. In Robert Heath, ed., *New Curricula,* 231–48. New York: Harper and Row, 1964.

Crowther, Bosley. "The Screen: 'Blackboard Jungle.'" *New York Times,* March 21, 1955, 21.

Crowther-Heyck, Hunter. *Herbert A. Simon: The Bounds of Reason in Modern America.* Baltimore: Johns Hopkins University Press, 2005.

Cuban, Larry. *How Teachers Taught: Constancy and Change in American Classrooms, 1880–1990.* 2nd ed. New York: Teachers College Press, 1993.

Dalmedico, Amy Dahan. "L'essor des mathématiques appliquées aux États-Unis." *Revue d'Histoire des Mathématiques* 2 (1996): 149–213.

———. "An Image Conflict in Mathematics after 1945." In Umberto Bottazini and Amy Dahan Dalmedico, eds., *Changing Images in Mathematics: From the French Revolution to the New Millennium,* 223–53. New York: Taylor and Francis, 2001.

Daston, Lorraine. *Classical Probability in the Enlightenment.* Princeton: Princeton University Press, 1988.

———. "Enlightenment Calculations." *Critical Inquiry* 21 (1994): 182–202.

Davie, George Elder. *The Democratic Intellect: Scotland and Her Universities in the Nineteenth Century.* 2nd ed. Edinburgh: Edinburgh University Press, 1964 [1961].

Davies, Gareth. *See Government Grow: Education Politics from Johnson to Reagan.* Lawrence: University Press of Kansas, 2007.

Davis, Robert B. *The Changing Curriculum: Mathematics.* Washington, DC: Association for Supervision and Curriculum Development, NEA, 1967.

Dear, Peter. *Discipline and Experience: The Mathematical Way in the Scientific Revolution.* Chicago: University of Chicago Press, 1995.

Define, Donald F. "Mathematics: Formulation of the Curriculum at Rich Township High School." *Clearing House* 36 (April 1962): 460–63.

DelFattore, Joan. *What Johnny Shouldn't Read: Textbook Censorship in America.* New Haven: Yale University Press, 1992.

DeMott, Benjamin. "The Math Wars." *American Scholar* 31 (1962): 296–310.

Dewey, John. *The Child and the Curriculum.* Chicago: University of Chicago Press, 1902.

———. *How We Think.* Boston: D. C. Heath, 1910.

———. "My Educational Creed." In J. Dewey, *Teachers Manuals*, 3–18. New York: E. L. Kellogg, 1899.

———. *The School and Society.* 2nd rev. ed. Chicago: University of Chicago Press, 1947 [lectures originally given 1899].

Diamond, Sara. *Not by Politics Alone: The Enduring Influence of the Christian Right.* New York: Guilford Press, 1998.

Dickens, Charles. *Hard Times for These Times.* New York: Penguin Books, 1995 [1854].

Dickson, Paul. *Sputnik: The Shock of the Century.* New York: Berkley Books, 2001.

Divine, Robert A. *The Sputnik Challenge.* New York: Oxford University Press, 1993.

Doherty, Thomas. *Cold War, Cool Medium: Television, McCarthyism, and American Culture.* New York: Columbia University Press, 2003.

Donahue, David. "Serving Students, Science, or Society? The Secondary School Physics Curriculum in the United States, 1930–1965." *History of Education Quarterly* 33 (1993): 321–52.

Donovan, Robert. "Killian Missile Czar: Ike Picks M.I.T. Head to Rush Research and Development." *Boston Daily Globe*, November 8, 1957, 1.

Dow, Peter B. *Schoolhouse Politics: Lessons from the Sputnik Era.* Cambridge: Harvard University Press, 1991.

Drake, Stillman, ed. and trans. *Discoveries and Opinions of Galileo.* New York: Anchor Books, 1957.

Dupre, Thomas Steven. "The University of Illinois Committee on School Mathematics and the 'New Mathematics' Controversy." Ph.D. diss., University of Illinois, 1986.

Duren, Peter, ed. *A Century of Mathematics in America*. 3 vols. Providence: American Mathematical Society, 1989.

Easterday, Kenneth. "Some Wrong Slants Righted concerning 'New' Math." *Alabama School Journal* 82 (January 1965): 4–5.

Edelson, Edward. "Reforming the Numbers Racket." *Washington Post*, February 25, 1973, BW10.

Education Policies Commission of the National Education Association, American Association of School Administrators, and the Executive Committee of the American Council on Education. *Education and National Security*. Washington, DC: Education Policies Commission and the American Council on Education, 1951.

Edwards, Paul N. *The Closed World: Computers and the Politics of Discourse in Cold War America*. Cambridge: MIT Press, 1996.

Eisenhower, Dwight D. "Farewell Radio and Television Address to the American People" [January 17, 1961]. In Carroll W. Pursell, Jr., ed., *The Military-Industrial Complex*, 204–8. New York: Harper and Row, 1972.

———. *The White House Years*, vol. 2: *Waging War and Peace, 1956–1961*. Garden City, NJ: Doubleday, 1965.

Elliott, David L. "Textbooks and the Curriculum in the Postwar Era: 1950–1980." In David L. Elliott and Arthur Woodward, eds., *Textbooks and Schooling in the United States: Eighty-ninth Yearbook of the National Society for the Study of Education*, 42–55. Chicago: University of Chicago Press, 1990.

Engelhardt, Tom. *The End of Victory Culture: Cold War America and the Disillusioning of a Generation*. Amherst: University of Massachusetts Press, 1995.

England, J. Merton. *A Patron for Pure Science: The National Science Foundation's Formative Years, 1945–57*. Washington, DC: National Science Foundation, 1982.

English, Raymond. "The Politics of Textbook Adoption." *Phi Delta Kappan* 62 (December 1980): 275–78.

Erickson, Paul. "Mathematical Models, Rational Choice, and the Search for Cold War Culture." *Isis* 101 (2010): 386–92.

Erickson, Paul, Judy L. Klein, Lorraine Daston, Rebecca Lemov, Thomas Sturm, and Michael D. Gordin. *How Reason Almost Lost Its Mind: The Strange Career of Cold War Rationality*. Chicago: University of Chicago Press, 2013.

Faber, Harold. "Curriculum Revision: A Continuing Process." *New York Times*, January 8, 1973.

"Family Zest for Learning." *Life*, April 21, 1958, 103–11.

Fang, Joong. *Bourbaki*. Hauppauge: Paideia Press, 1970.

——. *Numbers Racket: The Aftermath of the "New Math."* Port Washington, NY: Kennikat Press, 1968.

Fehr, Howard F. "The Administration of Mathematics Education in the United States of America." *School Science and Mathematics* 55 (1955): 340–44.

——. "Helping Our Teachers." *Mathematics Teacher* 50 (October 1957): 451–42.

——. "Modern Mathematics and Good Pedagogy." *Arithmetic Teacher* 10 (November 1963): 402–11.

Ferguson, W. Eugene. "Mathematics in Newton." *NAASP Bulletin* 52 (April 1968): 55–64.

Ferreirós, José. *Labyrinth of Thought: A History of Set Theory and Its Role in Modern Mathematics*. Basel: Birkhäuser Verlag, 1999.

Fey, James T., and Anna O. Graeber. "From the New Math to the *Agenda for Action*." In George M. A. Stanic and Jeremy Kilpatrick, eds., *A History of School Mathematics*, 1:521–58. Reston, VA: National Council of Teachers of Mathematics, 2003.

Feynman, Richard P. "New Textbooks for the New Mathematics." *California Education* 3 (1966): 8–14.

Fine, Benjamin. "Satellite Called Spur to Education: Soviet Success Shows Need for a Major U.S. Effort, College Heads Agree." *New York Times*, October 12, 1957, 3.

——. "Soviet Satellite Blow to Educators in U.S.: Fail to Understand Why Soviet Beat U.S." *Boston Herald*, October 11, 1957, 22.

Fiske, Edward B. "Suburban Schools Are Evolving 'Basic' Curriculums Geared to 1970's." *New York Times*, June 15, 1977, 45.

Fitzgerald, Frances. *America Revised: History Schoolbooks in the Twentieth Century*. Boston: Little, Brown, 1979.

——. "A Reporter at Large: A Disagreement in Baileyville." *New Yorker*, January 16, 1984, 47–90.

Fleck, Ludwik. *Genesis and Development of a Scientific Fact*. Ed. Thaddeus J. Treun and Robert K. Merton, trans. Fred Bradley and Thaddeus J. Treun. Chicago: University of Chicago Press, 1979.

Flesch, Rudolph. *Why Johnny Can't Read—And What You Can Do about It*. New York: Harper and Brothers, 1955.

Forman, Paul. "Behind Quantum Electronics: National Security as Basis for Physical Research in the United States, 1940–1960." *Historical Studies in the Physical and Biological Sciences* 18 (1987): 149–229.

Formisano, Ronald P. *Boston against Busing: Race, Class, and Ethnicity in the 1960s and 1970s*. Chapel Hill: University of North Carolina Press, 1991.

Fortune, Thomas. "The New Math: Adherents Multiplying Rapidly as the Years Add Up." *Los Angeles Times*, October 24, 1971, OC1.

———. "Schools Fighting to Regain Trust of Disenchanted Parents." *Los Angeles Times*, December 3, 1972, A1.

Foucault, Michel. "About the Beginning of the Hermeneutics of the Self: Two Lectures at Dartmouth" [1980]. *Political Theory* 21 (1993): 198–227.

Galison, Peter. "Structure of Crystal, Bucket of Dust," In Apostolos Doxiadis and Barry Mazur, eds., *Circles Disturbed: The Interplay of Mathematics and Narrative*, 52–78. Princeton: Princeton University Press, 2012.

Galison, Peter, and Bruce Hevly. *Big Science: The Growth of Large Scale Research*. Stanford: Stanford University Press, 1992.

Gallup, George H. "Ninth Annual Gallup Poll of the Public's Attitudes toward the Public Schools." *Phi Delta Kappan* 59 (September 1977): 33–36.

Gardner, John. *Excellence: Can We Be Equal and Excellent Too?* New York: Harper and Row, 1961.

Garigliano, Leonard J. "Arithmetic Computation Scores: Or Can Children in Modern Mathematics Program Really Compute?" *School Science and Mathematics* 75 (1975): 399–412.

Gerstle, Gary. "Race and the Myth of the Liberal Consensus." *Journal of American History* 82 (1995): 579–86.

Ghamari-Tabrizi, Sharon. *The Worlds of Herman Kahn: The Intuitive Science of Thermonuclear War*. Cambridge: Harvard University Press, 2005.

Gibb, E. Glenadine, Phillip S. Jones, and Charlotte W. Junge. "Number and Operation." In *The Growth of Mathematical Ideas: Grades K–12, Twenty-fourth Yearbook of NCTM*, 7–64. Washington, DC: National Council of Teachers of Mathematics, 1959.

Gibbs, Sam M. "New Math: Curriculum Changes." *Texas Outlook* 48 (April 1964): 34, 54.

Gilbert, James. *A Cycle of Outrage: America's Reaction to the Juvenile Delinquent in the 1950s*. New York: Oxford University Press, 1986.

Gilman, Nils. *Mandarins of the Future: Modernization Theory in Cold War America*. Baltimore: Johns Hopkins University Press, 2003.

Goffman, Erving. "On the Characteristics of Total Institutions" [1957]. In E. Goffman, *Asylums: Essays on the Social Situation of Mental Patients and Other Inmates*, new ed., 1–124. New Brunswick, NJ: Transaction, 2007 [1961].

Goldin, Claudia, and Lawrence Katz. "Human Capital and Social Capital: The Rise of Secondary Schooling in America, 1910–1940." *Journal of Interdisciplinary History* 29 (1999): 683–723.

———. *The Race between Education and Technology*. Cambridge: Belknap Press of Harvard University Press, 2008.

Goldstein, Jack S. *A Different Sort of Time: The Life of Jerrold R. Zacharias Scientist, Engineer, Educator*. Cambridge: MIT Press, 1992.

Goldstein, Jan. *The Post-Revolutionary Self: Politics and Psyche in France, 1750–1850*. Cambridge: Harvard University Press, 2005.

Golub, Adam Benjamin. "Into the Blackboard Jungle: Educational Debate and Cultural Change in 1950s America." Ph.D. diss., University of Texas at Austin, 2004.

Goodman, Jr., George. "Prof. Edward G. Begle, Chief Proponent of 'New Math.'" *New York Times*, March 3, 1978, B2.

Graham, Hugh Davis. *The Uncertain Triumph: Federal Education Policy in the Kennedy and Johnson Years*. Chapel Hill: University of North Carolina Press, 1984.

Grapí, Pere. "The Role of Chemistry Textbooks and Teaching Institutions in France at the Beginning of the Nineteenth Century in the Controversy about Berthollet's Chemical Affinities." In Peter Heering and Rolan Wittje, eds., *Learning by Doing: Experiments and Instruments in the History of Science Teaching*, 55–70. Stuttgart: Franz Steiner Verlag, 2011.

Grattan-Guinness, Ivor. *The Rainbow of Mathematics: A History of the Mathematical Sciences*. New York: Norton, 1997.

———. *The Search for Mathematical Roots, 1870–1940: Logics, Set Theories and the Foundation of Mathematics from Cantor through Russell to Gödel*. Princeton: Princeton University Press, 2000.

Gray, Jeremy J. *The Hilbert Challenge*. New York: Oxford University Press, 2000.

———. *János Bolyai, Non-Euclidean Geometry, and the Nature of Space*. Cambridge: MIT Press, 2004.

———. *Plato's Ghost: The Modernist Transformation of Mathematics*. Princeton: Princeton University Press, 2008.

Green, Larry. "'New Math' in New Guise: The Old Math." *Los Angeles Times*, November 28, 1977, A6, 18.

Greenberg, Daniel S. *The Politics of Pure Science*. New York: New American Library, 1967.

Greenspan, H. P. "Applied Mathematics as a Science." *American Mathematical Monthly* 68 (1961): 872–80.

Guthrie, James W. "A Political Case History: Passage of the ESEA." *Phi Delta Kappan* 49 (1968): 302–6.

Halberstam, David. *The Best and the Brightest*. New York: Random House, 1972.

Hall, Wendell. "Eugene Develops a Mathematics Program." *NAASP Bulletin* 52 (April 1968): 74–85.

Haney, David Paul. *The Americanization of Social Science: Intellectuals and Public Responsibility in the Postwar United States*. Philadelphia: Temple University Press, 2008.

Harnischfeger, Annegret, and David E. Wiley. "The Marrow of Achievement Test Score Declines." *Educational Technology* 16 (June 1976): 5–14.

Hart, Walter W., Veryl Schult, and Henry Swain. *First Year Algebra*. Boston: D. C. Heath, 1957.

Hartman, Andrew. *Education and the Cold War: The Battle for the American School.* New York: Palgrave Macmillan, 2008.

Haveman, Robert H. *Poverty Policy and Poverty Research: The Great Society and the Social Sciences.* Madison: University of Wisconsin Press, 1987.

Hawley, Newton S., and Patrick Suppes. "Geometry in the First Grade." In SMSG, *Report of a Conference on Elementary School Mathematics,* 36–37. New Haven: Yale University Press, 1959.

Hayden, Robert W. "A History of the 'New Math' Movement in the United States." Ph.D. diss., Iowa State University, 1981.

Heater, Derek. *A Brief History of Citizenship.* New York: New York University Press, 2004.

Heath, Thomas L., ed. *The Thirteen Books of Euclid's Elements: Translated from the Text of Heiberg.* 2nd rev. ed. 3 vols. Cambridge: Cambridge University Press, 1926.

Hechinger, Fred M. "The Back-to-the-Basics Impact." *Today's Education* 67 (February–March 1978): 31–32.

———. "Time + Trial = Acceptance." *New York Times,* August 6, 1972, E7.

———. "Where Have All the Innovations Gone?" *New York Times,* November 16, 1975, ED30.

Hefley, James C. *Are Textbooks Harming Your Children? Norma and Mel Gabler Take Action and Show You How.* Milford, MI: Mott Media, 1979.

Helgeson, Stanley L., Patricia E. Blosser, and Robert W. Howe. *The Status of Pre-College Science Mathematics, and Social Studies Practices in U.S. Schools: An Overview and Summaries of Three Studies.* SE 78–71. Washington, DC: U.S. Government Printing Office, 1978.

Herman, Ellen. *The Romance of American Psychology: Political Culture in the Age of Experts.* Berkeley: University of California Press, 1995.

Hilbert, David. *Foundations of Geometry.* Trans. Leo Unger. La Salle, IL: Open Court, 1971 [from material developed in 1898–99].

Hildebrandt, Martha Elizabeth. "Parental Attitudes about Mathematics Education: New Math in Winnetka." Ph.D. diss., Northwestern University, 1976.

Hirsch, Jr., E. D. "The Roots of the Education Wars." In Tom Loveless, ed., *The Great Curriculum Debate: How Should We Teach Reading and Math?* 13–24. Washington, DC: Brookings Institution Press, 2001.

Hodenfield, G. K. "What's New in the Curriculum." *PTA Magazine* 61 (January 1967): 7–9.

Hodgson, Godfrey. *America in Our Time.* Garden City, NJ: Doubleday, 1976.

Hofstadter, Richard. *Anti-Intellectualism in American Life.* New York: Vintage Books, 1963.

Hogben, Lancelot. *Mathematics for the Million: A Popular Self Educator.* London: George Allen and Unwin, 1936.

Hollinger, David A. "Science as Weapon in *Kulturekämpfe* in the United States during and after World War II." In D. Hollinger, *Science, Jews, and Secular Culture: Studies in Mid-Twentieth-Century American Intellectual History*, 155–74. Princeton: Princeton University Press, 1996.

Holsinger, Bruce. *The Premodern Condition: Medievalism and the Making of Theory*. Chicago: University of Chicago Press, 2005.

House Study Group of the Texas House of Representatives. "Texas Textbook Controversy." No. 101 (May 9, 1984).

"How They're Teaching Math to Your Kids." *Changing Times* 23 (November 1969): 19–22.

Howson, A. G., ed. *Developments in Mathematical Education: Proceedings of the Second International Conference on Mathematical Education*. Cambridge: Cambridge University Press, 1973.

Huettig, Alice, and John M. Newell. "Attitudes toward Introduction of Modern Mathematics Program by Teachers with Large and Small Number of Years' Experience." *Arithmetic Teacher* 13 (February 1966): 125–30.

Huff, Darrell. "Understanding the New Math?" *Harper's Magazine* 231 (September 1965): 134–37.

Hunter, Ian. "Assembling the School." In Andrew Barry, Thomas Osborne, and Nikolas Rose, eds., *Foucault and Political Reason: Liberalism, Neo-Liberalism, and the Rationalities of Government*, 143–66. Chicago: University of Chicago Press, 1996.

Hutchins, Robert. *The Conflict in Education in a Democratic Society*. New York: Harper and Brothers, 1953.

"Improvement in Arithmetical Instruction." *American Journal of Education* 2 (1827): 30–37.

"Inside Numbers." *Time*, January 31, 1964, 34–35.

Irons, Jerry. "Five Years after the Bomb: Or It's Time Again to Adopt New Elementary Math Books." *Texas Outlook* 54 (February 1970): 40.

Jackson, Kenneth. *Crabgrass Frontier: The Suburbanization of the United States*. New York: Oxford University Press, 1985.

Jackson, Philip W. *Life in Classrooms*. New York: Holt, Rinehart and Winston, 1968.

James, Glenn, ed. *Tree of Mathematics*. Pacoima, CA: Digest Press, 1957.

Jardini, David Raymond. "Out of the Blue Yonder: The RAND Corporation's Diversification into Social Welfare Research, 1946–1968." Ph.D. diss., Carnegie Mellon University, 1996.

Jeffrey, Julie Roy. *Education for Children of the Poor: A Study of the Origins and Implementation of the Elementary and Secondary Education Act of 1965*. Columbus: Ohio State University Press, 1978.

Jenkinson, Edward B. *Censors in the Classroom: The Mind Benders*. Carbondale: Southern Illinois University Press, 1979.

Johnson, Lyndon B. *The Vantage Point: Perspectives of the Presidency, 1963–1969*. New York: Holt, Rinehart, and Winston, 1971.

Johnson, Richard, Margaret Ullmann Miller, and Steven D. Kerr. *Elementary Algebra*. Menlo Park, CA: Benjamin/Cummings, 1981.

Jones, Matthew L. *The Good Life in the Scientific Revolution: Descartes, Pascal, Leibniz, and the Cultivation of Virtue*. Chicago: University of Chicago Press, 2006.

Kaestle, Carl F. *Pillars of the Republic: Common Schools and American Society, 1780–1860*. New York: Hill and Wang, 1983.

Kaiser, David. *American Physics and the Cold War Bubble*. Chicago: University of Chicago Press, forthcoming.

———. "Cold War Requisitions, Scientific Manpower, and the Production of American Physicists after World War II." *Historical Studies in the Physical Sciences* 33 (2002): 131–59.

———. *Drawing Theories Apart: The Dispersion of Feynman Diagrams in Postwar Physics*. Chicago: University of Chicago Press, 2005.

———. *How the Hippies Saved Physics: Science, Counterculture, and the Quantum Revival*. New York: W. W. Norton, 2011.

———. "A Mannheim for All Seasons: Bloor, Merton, and the Roots of the Sociology of Scientific Knowledge." *Science in Context* 11 (1998): 51–87.

———, ed. *Pedagogy and the Practice of Science: Historical and Contemporary Perspectives*. Cambridge: MIT Press, 2005.

———. "The Physics of Spin: Sputnik Politics and American Physicists in the 1950s." *Social Research* 73 (2006): 1225–52.

———. "The Postwar Suburbanization of American Physics." *American Quarterly* 56 (2004): 851–88.

Karabel, Jerome. *The Chosen: The Hidden History of Admission and Exclusion at Harvard, Yale, and Princeton*. New York: Houghton Mifflin, 2005.

Katz, Michael B. *The Irony of Early School Reform: Educational Innovation in Mid-Nineteenth Century Massachusetts*. Cambridge: Harvard University Press, 1968.

Kearns [Goodwin], Doris. *Lyndon Johnson and the American Dream*. New York: Harper and Row, 1976.

Kendig, Frank. "Does the New Math Add Up?" *New York Times*, January 6, 1974, E14, 28, 32.

Kevles, Daniel. "Cold War and Hot Physics: Science, Security, and the American State, 1945–1956." *Historical Studies in the Physical and Biological Sciences* 20 (1990): 239–64.

———. "The National Science Foundation and the Debate over Postwar Research Policy, 1942–1945: A Political Interpretation of *Science—The Endless Frontier*." *Isis* 68 (1977): 5–26.

Killian, Jr., James R. *Sputnik, Scientists, and Eisenhower: A Memoir of the First Special Assistant to the President for Science and Technology.* Cambridge: MIT Press, 1977.

Kilpatrick, Jeremy, and J. Fred Weaver. "The Place of William A. Brownell in Mathematics Education." *Journal for Research in Mathematics Education* 8 (November 1977): 382–84.

Kimmage, Michael. *The Conservative Turn: Lionel Trilling, Whittaker Chambers, and the Lessons of Anti-Communism.* Cambridge: Harvard University Press, 2009.

King, David C., and Zachary Karabell. *The Generation of Trust: How the U.S. Military Has Regained the Public's Confidence since Vietnam.* Washington, DC: AEI Press, 2003.

Klarman, Michael J. *Brown v. Board of Education and the Civil Rights Movement.* New York: Oxford University Press, 2007.

Kleinman, Daniel Lee. *Politics on the Endless Frontier: Postwar Research Policy in the United States.* Durham, NC: Duke University Press, 1995.

Kliebard, Herbert M., and Barry M. Franklin. "The Ascendance of Practical and Vocational Mathematics, 1893–1945: Academic Mathematics under Siege." In George M. A. Stanic and Jeremy Kilpatrick, eds., *A History of School Mathematics*, 1:399–440. Reston, VA: National Council of Teachers of Mathematics, 2003.

Kline, Morris. "Logic versus Pedagogy." *American Mathematical Monthly* 77 (March 1970): 264–82.

——. "Math Teaching Reforms Assailed as Peril to U.S. Scientific Progress." *New York University Alumni News* (October 1961): 1, 3, 8.

——. *Mathematical Thought from Ancient to Modern Times.* 3 vols. New York: Oxford University Press, 1972.

——. "Mathematics Texts and Teachers: A Tirade." *Mathematics Teacher* 49 (March 1956): 162–72.

——. *Why Johnny Can't Add: The Failure of the New Math.* New York: Vintage Books, 1973.

Knoll, Erwin. "Educators Told of School Need." *Washington Post*, October 12, 1957, C12.

Kolata, Gina Bari. "Aftermath of the New Math: Its Originators Defend It." *Science*, n.s. 195, no. 4281 (March 4, 1977): 854–57.

Kolesnik, Walter B. "Mental Discipline and the Modern Curriculum." *Peabody Journal of Education* 35 (March 1958): 298–303.

Kounine, Laura, John Marks, and Elizabeth Truss. "The Value of Mathematics." *Reform Report* (February 2008).

Kozol, Jonathan. *Death at an Early Age: The Destruction of the Hearts and Minds of Negro Children in the Boston Public Schools.* New York: Penguin, 1985 [1967].

Krieghbaum, Hillier, and Hugh Rawson. *An Investment in Knowledge: The First Dozen Years of the Science Foundation's Summer Institutes Programs to Improve Secondary School Science and Mathematics Teaching, 1954–1965.* New York: New York University Press, 1969.

Kripke, Saul A. *Wittgenstein on Rules and Private Language: An Elementary Exposition.* Oxford: Blackwell, 1982.

Kuhn, Thomas S. *The Structure of Scientific Revolutions.* 3rd ed. Chicago: University of Chicago Press, 1996 [1962].

Lagemann, Ellen Condliffe. *An Elusive Science: The Troubling History of Education Research.* Chicago: University of Chicago Press, 2000.

Lakatos, Imre. *Proofs and Refutations: The Logic of Mathematical Discovery.* Ed. John Worrall and Elie Zahar. Cambridge: Cambridge University Press, 1976.

Lankford, Jr., Francis G. "Implications of the Psychology of Learning for the Teaching of Mathematics." In *The Growth of Mathematical Ideas: Grades K–12, Twenty-Fourth Yearbook of NCTM,* 405–30. Washington, DC: National Council of Teachers of Mathematics, 1959.

Latham, Michael E. *Modernization as Ideology: American Social Science and "National Building" in the Kennedy Era.* Chapel Hill: University of North Carolina Press, 2000.

———. *The Right Kind of Revolution: Modernization, Development, and U.S. Foreign Policy from the Cold War to the Present.* Ithaca, NY: Cornell University Press, 2011.

Lave, Jean. *Cognition in Practice: Mind, Mathematics, and Culture in Everyday Life.* Cambridge: Cambridge University Press, 1988.

Lax, Peter. "The Flowering of Applied Mathematics in America." In Peter Duren, ed., *A Century of Mathematics in America,* 2:455–66. Providence: American Mathematical Society, 1989 [1976].

Lemann, Nicholas. *The Big Test: The Secret History of the American Meritocracy.* New York: Farrar, Straus and Giroux, 1999.

Lemov, Rebecca. *World as Laboratory: Experiments with Mice, Mazes, and Men.* New York: Hill and Wang, 2005.

Lerner, Daniel, with Lucille W. Pevsner. *The Passing of Traditional Society: Modernizing the Middle East.* Glencoe, IL: Free Press, 1958.

Lerner, Robert, Althea K. Nagai, and Stanley Rothman. *Molding the Good Citizen: The Politics of High School History Texts.* Westport, CT: Praeger, 1995.

Leslie, Stuart W. *The Cold War and American Science: The Military-Industrial-Academic Complex at MIT and Stanford.* New York: Columbia University Press, 1993.

"Lessons of the New Math." Editorial. *Wall Street Journal,* August 17, 1972, 10.

Light, Jennifer S. *From Warfare to Welfare: Defense Intellectual and Urban*

Problems in Cold War America. Baltimore: Johns Hopkins University Press, 2003.

Lilliston, Lynn. "Getting Back to Schooling Fundamentals." *Los Angeles Times*, January 30, 1974, E1, 4.

Lindsay, Frank B. "Mathematics Comes of Age in California High Schools." *California Journal of Secondary Education* 35 (February 1960): 79–83.

———. "Recent Developments in High School Mathematics in California." *Journal of Secondary Education* 36 (December 1961): 478–85.

Lockard, J. David., ed. *Third Report of the Information Clearinghouse on New Science and Mathematics Curricula*. American Association for the Advancement of Science and the Science Teaching Center, University of Maryland, March 1965.

Lomask, Milton. *A Minor Miracle: An Informal History of the National Science Foundation*. Washington, DC: National Science Foundation, 1976.

London, Herbert Ira. *Why Are They Lying to Our Children?* New York: Stein and Day, 1984.

Lora, Ronald. "Education: Schools as Crucible in Cold War America." In Robert H. Bremner and Gary W. Reichard, eds., *Reshaping America: Society and Institutions 1945–1960*, 223–60. Columbus: Ohio State University Press, 1982.

Loveless, Tom, ed. *The Great Curriculum Debate: How Should We Teach Reading and Math?* Washington, DC: Brookings Institution Press, 2001.

Lundgren, Anders, and Bernadette Bensaude-Vincent, eds. *Communicating Chemistry: Textbooks and Their Audiences, 1789–1939*. Canton, MA: Science History Publications, 2000.

MacLane, Saunders. "Mathematics at the University of Chicago: A Brief History." In Peter Duren, ed., *A Century of Mathematics in America*, 2:127–54. Providence: American Mathematical Society, 1989.

Maeroff, Gene I. "Issue and Debate: The Return to Fundamentals in the Nation's Schools." *New York Times*, December 6, 1975, 58.

Mallory, Virgil S. *Mathematics for Victory: An Emergency Course*. Chicago: Benj. H. Sanborn, 1943.

Marshall, J. Dan. "With a Little Help From Some Friends: Publishers, Protestors, and Texas Textbook Decisions." In Michael W. Apple and Linda K. Christian-Smith, eds., *The Politics of the Textbook*, 56–77. New York: Routledge, 1991.

Martin, Jay. *The Education of John Dewey: A Biography*. New York: Columbia University Press, 2002.

Martin, William. *With God on Our Side: The Rise of the Religious Right in America*. New York: Broadway, 1996.

Mashaal, Maurice. *Bourbaki: A Secret Society of Mathematicians*. Trans. Anna Pierrehumbert. Providence: American Mathematical Society, 2006.

"Math Textbook Is Target of B.H. [Beverly Hills] Parents." *Los Angeles Times*, November 30, 1972, WS5.

Matthews, Jay. "New Math Baffles Old Mathematician." *Washington Post*, November 15, 1972, A1, A13.

Matusow, Allen J. *The Unraveling of America: A History of Liberalism in the 1960s*. New York: Harper and Row, 1984.

McCulloch, Gary. "The Politics of the Secret Garden: Teachers and the School Curriculum in England and Wales." In Christopher Day, Alicia Fernandez, Trond E. Hauge, and Jorunn Moller, eds., *The Life and Work of Teachers: International Perspectives in Changing Times*, 26–37. London: Falmer Press, 1999.

McCurdy, Jack. "Back to 'Concrete' Addition, Subtraction: State Board Gives 'New Math' a 1–2 Punch." *Los Angeles Times*, May 10, 1974, A1.

——. "IQ Scores of California Pupils Drop for 6th Consecutive Year." *Los Angeles Times*, May 11, 1973, D1.

——. "Second Look: The New Math—Compounding Old Problems." *Los Angeles Times*, September 27, 1973, 1, 28–29.

McDougall, Walter A. *The Heavens and the Earth: A Political History of the Space Age*. New York: Basic Books, 1985.

McGirr, Lisa. *Suburban Warriors: The Origins of the New American Right*. Princeton: Princeton University Press, 2001.

McGuinn, Patrick J. *No Child Left behind and the Transformation of Federal Education Policy, 1965–2005*. Lawrence: University Press of Kansas, 2006.

McLellan, James A., and John Dewey. *The Psychology of Number and Its Applications to Methods of Teaching Arithmetic*. New York: D. Appleton, 1895.

McLeod, Douglas B. "From Consensus to Controversy: The Story of the NCTM Standards." In George M. A. Stanic and Jeremy Kilpatrick, eds., *A History of School Mathematics*, 1:753–818. Reston, VA: National Council of Teachers of Mathematics, 2003.

McLeod, Douglas B., Robert E. Stake, Bonnie P. Schappelle, Melissa Mellissinos, and Mark J. Gierl. "Setting the Standards: NCTM's Role in the Reform of Mathematics Education." In Senta A. Raizen and Edward D. Britton, eds., *Case Studies of U.S. Innovations in Mathematics Education*, 3:13–132. Dordrecht: Kluwer, 1996.

McMurrin, Sterling M. "The Curriculum and the Purposes of Education." In Robert Heath, ed., *New Curricula*, 262–84. New York: Harper and Row, 1964.

Meder, Jr., Albert E. "The Ancients versus the Moderns—A Reply." *Mathematics Teacher* 51 (October 1958): 428–33.

——. "Modern Mathematics and Its Place in the Secondary School." *Mathematics Teacher* 50 (October 1957): 418–23.

Menand, Louis. *The Marketplace of Ideas: Reform and Resistance in the American University*. New York: Norton, 2010.

Mendelsohn, Everett, Merritt Roe Smith, and Peter Weingart, eds. *Science Technology and the Military*. Dordrecht: Kluwer Academic, 1988.

Merrill, Richard J., and David W. Ridgway. *The CHEM Study Story*. San Francisco: W. H. Freeman, 1969.

Merton, Robert K. "Science and Democratic Social Structure" [1942]. In R. Merton, *Social Theory and Social Structure*, 550–61. Glencoe, IL: Free Press, 1957.

Meserve, Bruce E. *Fundamental Concepts of Algebra*. Cambridge, MA: Addison-Wesley, 1953.

———. "The University of Illinois List of Mathematical Competencies." *School Review* 61 (1953): 85–92.

Mialet, Hélène. *Hawking Incorporated: Stephen Hawking and the Anthropology of the Knowing Subject*. Chicago: University of Chicago Press, 2012.

Milam, Erika Lorraine. "Public Science of the Savage Mind: Contesting Cultural Anthropology in the Cold War Classroom." *Journal of the History of the Behavioral Sciences* 49 (2013): 306–30.

Mirowski, Philip. *Machine Dreams: Economics Becomes a Cyborg Science*. Cambridge: Cambridge University Press, 2002.

Mody, Cyrus C. M. "How I Learned to Stop Worrying and Love the Bomb, the Nuclear Reactor, the Computer, Ham Radio, and Recombinant DNA." *Historical Studies in the Natural Sciences* 38 (2008): 451–61.

Moffett, James. *Storm in the Mountains: A Case Study of Censorship, Conflict, and Consciousness*. Carbondale: Southern Illinois University Press, 1988.

Moon, Bob. *The "New Maths" Curriculum Controversy*. London: Falmer Press, 1986.

Moreau, Joseph. *Schoolbook Nation: Conflicts over American History Textbooks from the Civil War to the Present*. Ann Arbor: University of Michigan Press, 2003.

Morgan, Frank M., and Burnham L. Paige. *Algebra 1*. New York: Henry Holt, 1958.

Moss, Ruth. "Perplexing Merry-Go-Round: Does the New Math Add Up to an 'A' or an 'F'?" *Chicago Tribune*, May 7, 1973, B13.

Moulton, Jeanne. *How Do Teachers Use Textbooks? A Review of the Research Literature*. Washington, DC: U.S. Agency for International Development, 1997.

Moynihan, Daniel P. *Maximum Feasible Misunderstanding: Community Action in the War on Poverty*. New York: Free Press, 1969.

Mueller, Francis J. *Arithmetic: Its Structure and Concepts*. Englewood Cliffs: Prentice-Hall, 1956.

———. "Goals for School Mathematics? Educators and Parents Differ." *Virginia Journal of Education* 63 (January 1970): 21–22.

——. *Understanding the New Elementary School Mathematics: A Parent–Teacher Guide*. Belmont, CA: Dickenson Publishing, 1965.

Muller, Ian. "Mathematics and Education: Some Notes on the Platonic Program." *Apeiron* 24 (1991): 85–104.

Murdoch, John E. *Album of Science: Antiquity and the Middle Ages*. New York: Charles Scribner's Sons, 1984.

National Academy of Sciences—National Research Council. "Psychological Research in Education: Report of a Conference Sponsored by the Advisory Board on Education." Washington, DC: NAS-NRC, 1958.

National Commission on Excellence in Education. *A Nation at Risk: The Imperative for Educational Reform*. Washington, DC: U.S. Government Printing Office, 1983.

National Committee on Mathematical Requirements. *The Reorganization of Mathematics in Secondary Education*. [Oberlin, OH]: Mathematical Association of America, 1923.

National Congress of Parents and Teachers. *Proceedings: Annual Convention Official Reports and Records* 62 (May 18–21, 1958): 15–16.

National Council of Teachers of Mathematics (NCTM). *An Analysis of New Mathematics Programs*. Washington, DC: NCTM, 1963.

——. *Commission on Standards for School Mathematics*. Reston, VA: NCTM, 1989.

——."The Secondary Mathematics Curriculum: Report of the Secondary-School Committee of the National Council of Teachers of Mathematics." *Mathematics Teacher* 52 (May 1959): 389–417.

National Council of Teachers of Mathematics' Regional Orientation Conferences in Mathematics. *The Revolution in School Mathematics: A Challenge for Administrators and Teachers*. Washington, DC: NCTM, 1961.

National School Boards Association Research Report. *Back to Basics*. Report 1978–1. Washington, DC: NSBA, 1978.

National Science Foundation. *Annual Reports*. Washington, DC: Government Printing Office, 1951–58.

National Science Foundation, Science Curriculum Review Team. *Pre-College Science Curriculum Activities of the National Science Foundation*. Washington, DC: National Science Foundation, 1975.

Nelkin, Dorothy. *Science Textbook Controversies and the Politics of Equal Time*. Cambridge: MIT Press, 1977.

Nelson, Adam R. *The Elusive Ideal: Equal Educational Opportunity and the Federal Role in Boston's Public Schools, 1950–1985*. Chicago: University of Chicago Press, 2005.

Netz, Reviel. *The Shaping of Deduction in Greek Mathematics: A Study in Cognitive History*. Cambridge: Cambridge University Press, 1999.

"New Harvard Plan Backed by Faculty." *New York Times*, November 1, 1945, 25.

"New Math: A Story of Evolution." *Texas Outlook* 48 (April 1964): 33.

"The New Math: Does It Really Add Up?" *Newsweek*, May 10, 1965, 112–17.

"New Math Book Sales Add Up: Publishers Find a Profitable Equation in Texts as Teaching Is Revolutionized." *Business Week*, April 10, 1965, 117–20.

"The 'New' Mathematics." *California Education* 1 (November 1963): 16.

"New Math Fails Simple Tasks." *Los Angeles Sentinel*, March 29, 1973, B8.

"New Math Your Children May Be Learning." *Good Housekeeping* 160 (January 1965): 132–33.

New York State Commission on the Quality, Cost, and Financing of Elementary and Secondary Education. *The Fleischmann Report on the Quality, Cost, and Financing of Elementary and Secondary Education in New York State.* 3 vols. New York: Published by the author, 1972.

Newman, Frank. "The Era of Expertise: The Growth, The Spread and Ultimately the Decline of the National Commitment to the Concept of the Highly Trained Expert; 1945 to 1970." Ph.D. diss., Stanford University, 1981.

Ngai, Mae M. *Impossible Subjects: Illegal Aliens and the Making of Modern America.* Princeton: Princeton University Press, 2004.

Nickerson, Michelle M. *Mothers of Conservatism: Women and the Postwar Right.* Princeton: Princeton University Press, 2012.

Northrop, E. P. "Modern Mathematics and the Secondary School Curriculum." *Mathematics Teacher* 48 (October 1955): 386–93.

"Nostalgia's Child: Back to the Basics." Editor's page. *Phi Delta Kappan* 58 (March 1977): 521.

Nunley, B. G. "Programed Instruction: Shift to 'New Math.'" *Texas Outlook* 47 (September 1963): 29.

Nye, Mary Jo. "Historical Sources of Science-as-Social-Practice: Michael Polanyi's Berlin." *Historical Studies in the Physical and Biological Sciences* 37 (2007): 409–34.

O'Connor, Meg. "At '3-Rs' School on Southside, Goal Strictly Education." *Chicago Tribune*, September 19, 1976, 45.

Olesko, Kathryn M. *Physics as a Calling: Discipline and Practice in the Königsberg Seminar for Physics.* Ithaca, NY: Cornell University Press, 1991.

———. "Science Pedagogy as a Category of Historical Analysis: Past, Present, and Future." *Science and Education* 15 (2006): 863–80.

Oliver, Kendrick. *To Touch the Face of God: The Sacred, the Profane, and the American Space Program, 1957–1975.* Baltimore: Johns Hopkins University Press, 2012.

Olson, Richard. "Scottish Philosophy and Mathematics, 1750–1830." *Journal of the History of Ideas* 32 (1971): 29–44.

"On the Mathematics Curriculum of the High School." *American Mathematical Monthly* 69 (March 1962): 189–93.

Organization for European Economic Co-Operation. *New Thinking in School Mathematics.* 1961.

Orwell, George. *Nineteen Eighty-Four.* Ed. Bernard Crick. Oxford: Clarendon Press, 1984 [1949].

Osborne, Alan R., and F. Joe Crosswhite. "Forces and Issues Related to Curriculum and Instruction, 7–12." In *A History of Mathematics Education in the United States and Canada, Thirty Second Yearbook,* 155–300. Washington, DC: National Council of Teachers of Mathematics, 1970.

Osman, Suleiman. "The Decade of the Neighborhood." In Bruce J. Schulman and Julian E. Zelizer, eds., *Rightward Bound: Making America Conservative in the 1970s,* 106–27. Cambridge: Harvard University Press, 2008.

——. *The Invention of Brownstone Brooklyn: Gentrification and the Search for Authenticity in Postwar New York.* New York: Oxford University Press, 2011.

Ozmon, Howard, ed. *Contemporary Critics of Education.* Danville, IL: Interstate Printers and Publishers, 1970.

Packard, Vance. *The Status Seekers.* New York: Pocket Books, 1961 [1959].

Panofsky, Erwin. *Gothic Architecture and Scholasticism.* Wimmer Memorial Lecture, 1948. Latrobe, PA: St. Vincent Archabbey, 1951.

Parshall, Karen Hunger, and David E. Rowe. *The Emergence of the American Mathematical Research Community, 1876–1900: J. J. Sylvester, Felix Klein, and E. H. Moore.* Providence: American Mathematical Society, 1994.

Patterson, James T. *Brown v. Board of Education: A Civil Rights Milestone and Its Troubled Legacy.* New York: Oxford University Press, 2001.

——. *Restless Giant: The United States from Watergate to Bush v. Gore.* New York: Oxford University Press, 2005.

Payne, Joseph N. "The New Math and Its Aftermath, Grades K–8." In George M. A. Stanic and Jeremy Kilpatrick, eds., A History of School Mathematics, 1:559–98. Reston, VA: National Council of Teachers of Mathematics, 2003.

Peacock, Margaret. *Cold War Kids: Images of Soviet-American Childhood and the Collapse of Consensus, 1945–1968.* Chapel Hill: University of North Carolina Press, forthcoming.

Pells, Richard H. *The Liberal Mind in a Conservative Age: American Intellectuals in the 1940s and 1950s.* New York: Harper and Row, 1985.

Perkinson, Henry. *The Imperfect Panacea: American Faith in Education, 1865–1965.* New York: Random House, 1968.

Peterson, Iver. "Nation's Schools Renewing Stress on the Basics." *New York Times,* March 3, 1975, 1, 38.

Phillips, Harry L., and Marguerite Kluttz, *Modern Mathematics and Your Child.* Washington, DC: U.S. Government Printing Office, 1965.

Phillips-Fein, Kim. "Conservatism: A State of the Field." *Journal of American History* 98 (2011): 723–43.

Piaget, Jean. "Comments on Mathematical Education." In A. G. Howson, ed., *Developments in Mathematical Education: Proceedings of the Second International Conference on Mathematical Education*, 79–87. Cambridge: Cambridge University Press, 1973.

———. "How Children Form Mathematical Concepts." *Scientific American* 189, no. 5 (November 1953): 74–79.

———. "Science of Education and the Psychology of the Child" [1965]. In Howard E. Gruber and J. Jacques Vonèche, eds., *The Essential Piaget*, 695–725. London: Routledge and Kegan Paul, 1977.

———. *Le structuralisme*. Paris: Presses Universitaires de France, 1968.

Pike, Nicolas. *New and Complete System of Arithmetick, Composed for Use of Citizens of the United States*. 7th ed. Ed. Nathanial Lord. Boston: Thomas and Andrews, 1809.

Pines, Burton Yale. *Back to Basics: The Traditionalist Movement That Is Sweeping Grass-Roots America*. New York: William Morrow, 1982.

Plant, Rebecca Jo. *Mom: The Transformation of Motherhood in Modern America*. Chicago: University of Chicago Press, 2010.

Plato. *Meno*. In Edith Hamilton and Huntington Cairns, eds., and W. K. C. Guthrie, trans., *The Collected Dialogues of Plato including the Letters*, 353–84. Princeton: Princeton University Press, 1961.

Polanyi, Michael. *Science, Faith and Society*. London: Oxford University Press, 1946.

Pólya, George. *How to Solve It: A New Aspect of Mathematical Method*. Princeton: Princeton University Press, 1945.

———. *Induction and Analogy in Mathematics*. 2 vols. Princeton: Princeton University Press, 1954.

Porter, Theodore M. *Trust in Numbers: The Pursuit of Objectivity in Science and Public Life*. Princeton: Princeton University Press, 1995.

"Position Statements on Basic Skills." *Mathematics Teacher* 71 (February 1978): 147–55.

Price, Derek J. de Solla. *Little Science, Big Science*. New York: Columbia University Press, 1963.

Price, G. Baley. "Mathematics and Education." In Lee C. Deighton, ed., *The Encyclopedia of Education*, 6:81–90. New York: Macmillan, 1971.

"The Price of Mathophobia." *Time*, March 17, 1967, 61.

Quick, Suzanne Kay. "Secondary Impacts of the Curriculum Reform Movement: A Longitudinal Study of the Incorporation of Innovations of the Curriculum Reform Movement into Commercially Developed Curriculum Programs." Ph.D. diss., Stanford University, 1978.

Rafferty, Max. *On Education*. New York: Devin-Adair, 1968.

Rand, Ayn. *Atlas Shrugged*. New York: Random House, 1957.

Rappaport, David. "The Nuffield Mathematics Project." *Elementary School Journal* 71 (1971): 295–308.

Ravitch, Diane. *The Language Police: How Pressure Groups Restrict What Students Learn*. New York: Alfred A. Knopf, 2003.

———. *National Standards in American Education: A Citizen's Guide*. Washington, DC: Brookings Institution Press, 1995.

———. *The Troubled Crusade: American Education, 1945–1980*. New York: Basic Books, 1983.

Reardon, B. R. "I'm for the Modern Math. Here's Why." *Alabama School Journal* 83 (January 1966): 9.

Redding, M. Frank. *Revolution in the Textbook Publishing Industry*. Washington, DC: National Education Association, 1963.

Reese, William J. *America's Public Schools: From the Common School to "No Child Left Behind."* Baltimore: Johns Hopkins University Press, 2005.

Reider, Jonathan. "The Rise of the 'Silent Majority.'" In Steve Fraser and Gary Gerstle, eds., *The Rise and Fall of the New Deal Order, 1930–1980*, 243–68. Princeton: Princeton University Press, 1989.

Reuben, Julie A. *The Making of the Modern University: Intellectual Transformation and the Marginalization of Morality*. Chicago: University of Chicago Press, 1996.

Reys, Robert E., R. D. Kerr, and John W. Alspaugh. "Mathematics Curriculum Change in Missouri Secondary Schools." *School and Community* 56 (December 1969): 6–7, 9.

Richards, Joan L. "Historical Mathematics in the French Eighteenth Century." *Isis* 97 (2006): 700–713.

———. *Mathematical Visions: The Pursuit of Geometry in Victorian England*. Boston: Academic Press, 1988.

Rickover, H. G. *Education and Freedom*. New York: Dutton, 1959.

Riesman, David, with Reuel Denney and Nathan Glazer. *The Lonely Crowd: A Study of the Changing American Character*. New Haven: Yale University Press, 1950.

Rinker, Floyd. "Subject Matter, Students, Teachers, Methods of Teaching, and Space are Redeployed in the Newton, Massachusetts, High School." *Bulletin of the National Association of Secondary-School Principals* 42 (January 1958): 69–80.

Roberts, David Lindsay. *American Mathematicians as Educators, 1893–1923: Historical Roots of the "Math Wars."* Chestnut Hill, MA: Docent Press, 2012.

———. "The BKPS Letter of 1962: The History of a 'New Math' Episode." *Notices of the AMS* 51 (2004): 1062–63.

———. "Mathematicians in the Schools: The 'New Math' as an Arena of Professional Struggle, 1950–1970." Talk given at the History of Science Annual Meeting, November 20, 2004.

Roberts, David L., and Angela Walmsley. "The Original New Math: Storytelling versus History." *Mathematics Teacher* 96 (2003): 468–73.

Robin, Ron. *The Making of the Cold War Enemy*. Princeton: Princeton University Press, 2001.

Rockefeller Brothers Fund. "The Pursuit of Excellence: Education and the Future of America." Special Studies Project Report V, American at Mid-Century Series. Garden City, NJ: Doubleday, 1958.

Rodgers, Daniel T. *Age of Fracture*. Cambridge: Harvard University Press, 2011.

Rolland, Alvin E. "Making the Switch to Modern Mathematics." *School and Community* 48 (May 1962): 29.

Rose, Nikolas. *Governing the Soul: The Shaping of the Private Self.* 2nd ed. London: Free Association Books, 1999.

Rosenbloom, Paul C. "Mathematics in the World of Children." In Thomas L. Saaty and F. Joachim Weyl, eds., *The Spirit and the Uses of the Mathematical Sciences*, 68–104. New York: McGraw-Hill, 1969.

———, ed. *Modern Viewpoints in the Curriculum: National Conference on Curriculum Experimentation*. New York: McGraw-Hill, 1964.

———. "Teaching Mathematics to Gifted Students." In SMSG, *Report of a Conference on Elementary School Mathematics*, 33–35. New Haven: Yale University Press, 1959.

———. "What Is Coming in Elementary Mathematics." *Educational Leadership* 18 (November 1960): 96–100.

Rudolph, John. "From World War to Woods Hole: The Impact of Wartime Research Models on Curriculum Reform." *Teachers College Record* 104 (2002): 212–41.

———. *Scientists in the Classroom: The Cold War Reconstruction of American Science Education*. New York: Palgrave, 2002.

———. "Teaching Materials and the Fate of Dynamic Biology in American Classrooms after Sputnik." *Technology and Culture* 53 (2012): 1–36.

Ruff, Thomas P., and Donald C. Orlich. "How Do Elementary-School Principals Learn about Curriculum Innovations?" *Elementary School Journal* 74 (April 1974): 389–92.

Saaty, Thomas L. and F. Joachim Weyl, eds. *The Spirit and the Uses of the Mathematical Sciences*. New York: McGraw-Hill, 1969.

Sarason, Seymour B. *The Culture of the School and the Problem of Change*. Boston: Allyn and Bacon, 1971.

School Mathematics Study Group (SMSG). *Mathematics for High School: First Course in Algebra: Student's Text*. Rev. ed. New Haven: Yale University Press, 1961.

———. *Mathematics for High School: First Course in Algebra: Teacher's Commentary*. Rev. ed. New Haven: Yale University Press, 1961.

———. *Mathematics for High School: Geometry*. Rev. ed. New Haven: Yale University Press, 1961.

———. *Mathematics for High School: Geometry: Teacher's Commentary*. Rev. ed. New Haven: Yale University Press, 1961.

———. *Mathematics for Junior High School: Student's Text*. 2 vols. Rev. ed. New Haven: Yale University Press, 1961.

———. *Mathematics for Junior High School: Teacher's Commentary*. 2 vols. Rev. ed. New Haven: Yale University Press, 1961.

———. *Mathematics for the Elementary School, Book 1*. Palo Alto: Stanford University Press, 1965.

———. *Mathematics for the Elementary School, Book 1, Unit 53: Teacher's Commentary*. Palo Alto: Stanford University Press, 1965.

———. *Report of a Conference on Elementary School Mathematics*. New Haven: Yale University Press, 1959.

———. *Report of an Orientation Conference for Geometry with Coordinates, Chicago, Illinois, September 23, 1961*. Palo Alto: Stanford University Press, 1962.

——— [Max S. Bell, William G. Chinn, Mary McDermott, Richard S. Pieters, and Margaret Willerding]. *Studies in Mathematics, vol. 9: A Brief Course in Mathematics for Elementary School Teachers*. Rev. ed. Palo Alto: Stanford University Press, 1963.

———. *Studies in Mathematics, vol. 13: Inservice Course in Mathematics for Primary School Teachers*. Rev. ed. Palo Alto: Stanford University Press, 1965–66.

Schorling, Raleigh, Rolland R. Smith, and John R. Clark. *Algebra: First Course*. New York: World Book Company, 1949.

Schrecker, Ellen. *No Ivory Tower: McCarthyism and the Universities*. New York: Oxford University Press, 1986.

Schulman, Bruce J., and Julian E. Zelizer, eds. *Rightward Bound: Making America Conservative in the 1970s*. Cambridge: Harvard University Press, 2008.

Schwartz, Harry. "Dethroning the 'New Math.'" *New York Times*, July 20, 1973, 29.

———. "The New Math Faces a Counter-Revolution." *New York Times*, January 8, 1973, 85.

———. "The New Math Is Replacing Third 'R.'" *New York Times*, January 25, 1965, 18.

———. "'New Math' Leader Sees Peril in Haste." *New York Times*, December 31, 1964, 1, 16.

Seeley, Cathy L. "Mathematics Textbook Adoption in the United States." In

George M. A. Stanic and Jeremy Kilpatrick, eds., *A History of School Mathematics*, 2: 947–88. Reston, VA: National Council of Teachers of Mathematics, 2003.

Seth, Suman. *Crafting the Quantum: Arnold Sommerfield and the Practice of Theory, 1890–1926*. Cambridge: MIT Press, 2010.

Shapin, Steven. "The Philosopher and the Chicken: On the Dietetics of Disembodied Knowledge." In Christopher Lawrence and Steven Shapin, eds., *Science Incarnate: Historical Embodiments of Natural Knowledge*, 21–50. Chicago: University of Chicago Press, 1998.

———. *The Scientific Life: A Moral History of a Late Modern Vocation*. Chicago: University of Chicago Press, 2008.

———. *A Social History of Truth: Civility and Science in Seventeenth-Century England*. Chicago: University of Chicago Press, 1994.

Shapin, Steven, and Barry Barnes. "Head and Hand: Rhetorical Resources in British Pedagogical Writing, 1770–1850." *Oxford Review of Education* 2 (1976): 231–54.

Shapiro, Adam. "State Regulation of the Textbook Industry." In Adam R. Nelson and John L. Rudolph, eds., *Education and the Culture of Print in Modern America*, 173–90. Madison: University of Wisconsin Press, 2010.

Shapiro, Stewart. *Philosophy of Mathematics: Structure and Ontology*. New York: Oxford University Press, 1997.

Shute, William O., William E. Kline, William W. Shirk, and Leroy M. Willson. *Elementary Algebra*. New York: American Book Company, 1956.

Smith, Gilbert E. *The Limits of Reform: Politics and Federal Aid to Education, 1937–1950*. Updated ed. New York: Longman, 1989.

Smith, John F. "Case History, Charlotte-Mecklenburg." *NAASP Bulletin* 52 (April 1968): 65–73.

Smith, Louis M., and William Geoffrey. *The Complexities of an Urban Classroom: An Analysis toward a General Theory of Teaching*. New York: Holt, Rinehart and Winston, 1968.

Smith, R. W. "Letters to the Editor: Culturally It Adds Up." *Washington Post*, December 8, 1972, A27.

Smyth, Henry D. "The Stockpiling and Rationing of Scientific Manpower." *Bulletin of the Atomic Scientists* 7 (February 1951): 38–42, 64.

Snader, Daniel W. *Algebra: Meaning and Mastery, Book 1*. Philadelphia: John C. Winston Company, 1949.

Snider, Robert C. "Back to the Basics?" NEA Position Paper (August 1978), 6–9.

Société mathématique de France. *Structures algébraiques et structures topologiques*. Geneva: Enseignement mathématique, 1958.

Solovey, Mark, and Hamilton Cravens, eds. *Cold War Social Science: Knowledge Production, Liberal Democracy, and Human Nature*. New York: Palgrave Macmillan, 2012.

Spring, Joel. *The American School: 1642–1993*. 3rd ed. New York: McGraw-Hill, 1994.

Sproull, Lee Stephen Weiner, and David Wolf. *Organizing an Anarchy: Belief, Bureaucracy, and Politics in the National Institute of Education*. Chicago: University of Chicago Press, 1978.

Stake, Robert E., and Jack A. Easley, Jr. *Case Studies in Science Education*, vol. 1: *The Case Reports*. Washington, DC: Government Printing Office, 1978.

———. *Case Studies in Science Education*, vol. 2: *Design, Overview and General Findings*. Washington, DC: U.S. Government Printing Office, 1978.

Stanic, George M. A., and Jeremy Kilpatrick, eds. *A History of School Mathematics*. 2 vols. Reston, VA: National Council of Teachers of Mathematics, 2003.

Steelman, John R. *Science and Public Policy*, vol. 4: *Manpower for Research*. Washington, DC: Government Printing Office, 1947.

Stewart, Mac A., ed. *The Promise of Justice: Essays on Brown v. Board of Education*. Columbus: Ohio State University Press, 2008.

Stone, Marshall H. "Fundamental Issues in the Teaching of Elementary School Mathematics." In SMSG, *Report of a Conference on Elementary School Mathematics*, 7–8. New Haven: Yale University Press, 1959.

———. "Reminiscences of Mathematics at Chicago." In Peter Duren, ed., *A Century of Mathematics in America*, 2:183–90. Providence: American Mathematical Society, 1989 [1976].

———. "The Revolution in Mathematics." *American Mathematical Monthly* 68 (1961): 715–34.

Strecker, Edward A. *Their Mothers' Sons: The Psychiatrist Examines an American Problem*. New York: J. B. Lippincott, 1946.

Suppes, Patrick. "Adding Up the New Math." *PTA Magazine* 60 (October 1965): 8–10.

Sweeney, Daniel. "Why Mohole Was No Hole." *American Heritage of Invention and Technology* 9 (1993): 54–63.

Taylor, Charles. *Sources of the Self: The Making of the Modern Identity*. Cambridge: Harvard University Press, 1989.

Taylor, Ross. "NACOME: Implications for Teaching K–12." *Mathematics Teacher* 69 (October 1976): 458–63.

Tebbel, John William. *A History of Book Publishing in the United States*. 4 vols. New York: R. R. Bowker, 1972–81.

Thernstrom, Stephen. *Poverty and Progress: Social Mobility in a Nineteenth-Century City*. Cambridge: Harvard University Press, 1981.

Thom, René. "'Modern' Mathematics: An Educational and Philosophic Error." *American Scientist* 59 (1971): 695–99.

———. "Modern Mathematics: Does It Exist?" In A. G. Howson, ed., *Developments in Mathematical Education: Proceedings of the Second International*

Conference on Mathematical Education, 194–209. Cambridge: Cambridge University Press, 1973.

Thomas, Helen. "Julie: Learning the New Math." *Washington Post*, January 14, 1971, G6.

Thompson, Kenneth W., ed. *The Presidency and Education*. Lanham, MD: University Press of America, 1990.

Thorpe, Charles. *Oppenheimer: The Tragic Intellect*. Chicago: University of Chicago Press, 2006.

Tolkan, Constance H. "A Parent Looks at the New Math." *Independent School Bulletin* 31 (October 1971): 53–55.

Toth, Robert C. "Teaching of the 'New Math' Stirs Wide Debate among Teachers." *New York Times*, September 21, 1962, 31.

"Tryouts for Good Ideas." *Life*, April 14, 1958, 117–25.

Tyack, David, and Larry Cuban. *Tinkering toward Utopia: A Century of Public School Reform*. Cambridge: Harvard University Press, 1995.

Tyack, David, and Elisabeth Hansot. *Managers of Virtue: Public School Leadership in America, 1820–1980*. New York: Basic Books, 1982.

Tyler, Ralph W. *The Florida Accountability Program: An Evaluation of Its Educational Soundness and Implementation*. Washington, DC: National Education Association, 1978.

"An Underdog Profession Imperils the Schools." *Life*, March 31, 1958, 93–101.

Urban, Wayne J. *More Than Science and Sputnik: The National Defense Education Act of 1958*. Tuscaloosa: University of Alabama Press, 2010.

U.S. Department of Health, Education, and Welfare, Division of Educational Statistics and Bureau of Educational Research and Development. *Digest of Educational Statistics*. Bulletin 1963, no. 43. Washington, DC: U.S. Government Printing Office, 1963.

"U.S. High School." *Life*, April 22, 1946, 87–93.

Usiskin, Zalman. "A Personal History of the UCSMP Secondary School Curriculum, 1960–1999." In George M. A. Stanic and Jeremy Kilpatrick, eds., *A History of School Mathematics*, 1:673–736. Reston, VA: National Council of Teachers of Mathematics, 2003.

Veblen, Oswald. "A System of Axioms for Geometry." *Transactions of the American Mathematical Society* 5 (1904): 343–84.

Veysey, Laurence R. *The Emergence of the American University*. Chicago: University of Chicago Press, 1965.

Wagner, John. "The Objectives and Activities of the School Mathematics Study Group." *Mathematics Teacher* 53 (October 1960): 454–59.

Walmsley, Angela L. E. "A History of the 'New Mathematics' Movement and Its Relationship with the National Council of Teachers of Mathematics Standards." Ph.D. diss., Saint Louis University, 2001.

Wang, Jessica. *American Science in an Age of Anxiety: Scientists, Anticommu-

nism, and the Cold War. Chapel Hill: University of North Carolina Press, 1999.

Wang, Zuoyue. *In Sputnik's Shadow: The President's Science Advisory Committee and Cold War America.* New Brunswick, NJ: Rutgers University Press, 2008.

Warwick, Andrew. *Masters of Theory: Cambridge and the Rise of Mathematical Physics.* Chicago: University of Chicago Press, 2003.

Warwick, Andrew, and David Kaiser. "Conclusion: Kuhn, Foucault, and the Power of Pedagogy," In David Kaiser, ed., *Pedagogy and the Practice of Science: Historical and Contemporary Perspectives,* 393–409. Cambridge: MIT Press, 2005.

"The Waste of Fine Minds." *Life,* April 7, 1958, 89–97.

Waterman, Alan T. "Role of the Federal Government in Science Education." *Scientific Monthly* 82 (June 1956): 286–93.

Weart, Spencer R. *The Rise of Nuclear Fear.* Cambridge: Harvard University Press, 2012.

Weiss, Iris R. *Report of the 1977 National Survey of Science, Mathematics, and Social Studies Education SE 78–72.* Washington, DC: Government Printing Office, 1978.

Westbrook, Robert B. *John Dewey and American Democracy.* Ithaca, NY: Cornell University Press, 1991.

"What Makes Them Good." *Time,* October 21, 1957, 52.

"What the Schools Cannot Do." *Time,* April 16, 1973, 78–85.

"When You Get Back to Basics, You Get Back to Ford." *New York Times,* November 30, 1971, 41.

Whitfield, Stephen J. *The Culture of the Cold War.* 2nd ed. Baltimore: Johns Hopkins University Press, 1996.

"Why Johnny Can't Add." *Newsweek,* September 24, 1979, 72.

Whyte, William. *The Organization Man.* New York: Simon and Schuster, 1956.

Williams, Raymond. *Keywords: A Vocabulary of Culture and Society.* New ed. New York: Oxford University Press, 1983 [1976].

Willoughby, Stephen S. "What Is the New Mathematics." *NAASP Bulletin* 52 (April 1968): 4–15.

Wilson, Sloan. "It's Time to Close Our Carnival." *Life* 44 (March 24, 1958): 36–37.

———. *The Man in a Gray Flannel Suit.* New York: Simon and Schuster, 1955.

Wilson, Suzanne M. *California Dreaming: Reforming Mathematics Education.* New Haven: Yale University Press, 2003.

Wittgenstein, Ludwig. *Lectures on the Foundations of Mathematics, Cambridge 1939.* Ed. Cora Diamond. Chicago: University of Chicago Press, 1976.

———. *Philosophical Investigations.* Trans. G. E. M. Anscombe. Oxford: Basil Blackwell, 1968.

———. *Remarks on the Foundations of Mathematics.* Rev. ed. Ed. G. H. Von Wright, R. Rhees, G. E. M. Anscombe, and trans. G. E. M. Anscombe. Cambridge: MIT Press, 1978.

Wolfe, Audra. *Competing with the Soviets: Science, Technology, and the State in Cold War America.* Baltimore: Johns Hopkins University Press, 2013.

Woods, Peter, and Martyn Hammersley, eds. *School Experience: Explorations in the Sociology of Education.* New York: St. Martin's Press, 1977.

Wooton, William. *SMSG: The Making of a Curriculum.* New Haven: Yale University Press, 1965.

Wright, D. Franklin, and Bill D. New. *Introductory Algebra.* Boston: Allyn and Bacon, 1981.

Wylie, Philip. *Generation of Vipers.* New ed. New York: Pocket Books, 1955.

Yin, Robert K., and Douglas Yates. *Street-Level Governments: Assessing Decentralization and Urban Services (An Evaluation of Policy Research).* Santa Monica: RAND, 1974.

Zant, James H. "Better Mathematics Teaching with Special Reference to Oklahoma." *Journal of Secondary Education* 39 (April 1964): 188–92.

Zelizer, Julian E. "Rethinking the History of American Conservatism." *Reviews in American History* 38 (2010): 367–92.

Zilversmit, Arthur. *Changing Schools: Progressive Education Theory and Practice, 1930–1960.* Chicago: University of Chicago Press, 1993.

Zimmerman, Jonathan. *Whose America? Culture Wars in the Public Schools.* Cambridge: Harvard University Press, 2002.

Index

Albert, A. Adrian, 41, 52
algebra, 14, 61–62, 64–67
American Mathematical Society. *See under* mathematicians
applied mathematics. *See under* mathematics
arithmetic: certainty of, 18–19; in curriculum, 14, 76–77; difficulty of teaching, 76–77, 79–80, 106–7; modular, 55–56; as psychological problem, 86–92; in SMSG's textbooks, 53–59, 82–86

"back to basics": as conservative, 20, 139–41; as decentralized, 136; media coverage of, 135–36; politics of, 141–42
Beberman, Max, 15–16, 100, 109
Begle, Edward: and commercial publishers, 103; and curriculum evaluation, 126–27; and direction of SMSG, 2–3, 49, 80–82; on "modern" mathematics, 15; photograph of, 43; psychology, views of, 87, 90; reasons for initiating reform, 42, 113–14; and secondary school teachers, 98, 133; and standards, 133
Bers, Lipman, 70–71
Bestor, Arthur, 30–31, 75, 79, 137
Bourbaki, Nicolas, 18, 50–53, 67, 71–72, 94
Bourdieu, Pierre, 10, 132
Brauer, Richard, 42, 49
Bronk, Detlev, 39, 45, 59, 61
Brownell, William, 89–90
Bruner, Jerome, 87–90, 132
Bush, George H. W., 148
Bush, Vannevar, 26

Cambridge Conference on School Mathematics, 16
Colburn, Warren, 59–60
Cold War: and authoritarian personalities, 77–80; conformity in, 33; and science, 11–12, 14, 78–79, 93–94. *See also under* discipline, mental
College Entrance Examination Board (CEEB), 15, 25, 49–50, 66–67
Conant, James Bryant, 29, 109
Courant, Richard, 67–69, 72

Dewey, John, 28–29, 86–87
DeWitt, Nicolas, 37–38
Dieudonné, Jean, 51–52
discipline, mental: and Cold War, 22–35, 78–80, 115–19; and curriculum, 9–10, 20–21, 142–44; and McCarthyism, 30–31; and "transfer of training," 88–89. *See also under* mathematics
Dolciani, Mary P., 103–4, 146
Doolittle, James H., 26–27

education: bureaucracy within, 12, 96; Common Core, 9, 148; as decentralized, 24; of democratic citizens, 32–35; and historians, 149; legislation, 22, 24–25, 46; progressive education, 18, 28–32; and social reform 12–13, 25–26, 128–29, 131–33; and standards, 133, 141, 147–48; as universal, 11
Elementary and Secondary Education Act (1965), 118, 133
Eisenhower, Dwight, 23, 25, 38–40, 134

Fehr, Howard, 52, 77–78, 80, 99, 109, 122
Foucault, Michel, 9–10

Gabler, Mel and Norma, 139–40, 144
Gardner, John, 26, 34
geometry: in curriculum, 14; Euclidean,
    6–7, 51–52, 62; and intuition, 63–64; in
    SMSG's textbooks, 61–64
Gleason, Andrew, 52
Great Society, 97, 117–20, 129, 139

habits, mental. *See* discipline, mental
Hamilton, William, 8
Hutchins, Robert, 29–31

Johnson, Lyndon, 26, 117–20, 139

Kennedy, John F., 117–18
Kline, Morris, 20, 69–74, 124–27, 136,
    138–39

Lehrer, Tom, 1–2
literacy, 20–21

manpower, scientific. *See under* National
    Science Foundation
mathematicians: and American Mathe-
    matical Society, 41–42, 48–49, 67, 69;
    opposition to new math, 18, 47–48,
    67–69. *See also names of individual
    mathematicians*
mathematics: applied, 67–69; of bases, 1;
    as calculation, 5, 13–15, 54–55, 76, 137;
    as general reasoning, 2, 60–61, 91–92,
    114–16, 118–20; history of, 6–8, 14, 71–
    73; and induction, 93–94; as mental dis-
    cipline, 4–9, 12, 44–46, 59–61, 136–38,
    148–49; as "modern," 4, 74, 95, 113,
    115–17, 145; nature of, 5–6, 13–15, 49–
    50, 72–74; of sets, 1, 14, 19, 82–86, 94;
    as structure of systems, 14, 50–59, 63–
    68, 82–89, 94, 143–44; as technology of
    the self, 9–10; understanding of, com-
    pared to rote learning of, 15, 58–59, 66,
    124, 134–35, 146–47; unity of, 14, 62.
    See also algebra; arithmetic; Bourbaki,
    Nicolas; geometry; mathematicians
Meder, Albert E., Jr., 73–74, 98, 103–4
military-industrial complex, 3, 39–40

Moise, Edwin E., 103–4, 146
Montucla, Jean-Étienne, 7–8
Moynihan, Daniel Patrick, 129–30

National Council of Teachers of Mathe-
    matics (NCTM), 49, 71, 77, 98–99, 115,
    136, 146–48
National Defense Education Act (1958),
    40–41, 118, 133
National Science Foundation (NSF):
    Course Content Improvement Program
    of, 39–41, 80, 117, 134; evaluation of
    projects, 128, 137–38, 142–43; faith in,
    130–31; funding of, 36–41; and scien-
    tific manpower, 35–38, 118, 120
new math: as bipartisan, 40–41, 118; and
    citizenship, 11–12, 19; as coherent en-
    tity, 2, 17; extent of, 122–23; as fail-
    ure or success, 5, 96, 121, 145–46; in
    high schools, compared to elementary
    schools, 19, 96–97, 105, 111, 119–20;
    as international, 12; irony of, 94–95,
    119–20, 138–39; legacy of, 145–48; me-
    dia coverage of, 107, 124–25; oppo-
    nents, views of, 19–20, 124–25, 127, 131,
    134–44; other reforms, compared to, 48,
    149; and test scores, 97, 124–27, 146–
    47; and textbook adoption lists, 110–
    12. *See also* School Mathematics Study
    Group
"New Math" (Lehrer), 1
Nixon, Richard, 124, 129–30, 134

Orwell, George, 75–76, 79

Panofsky, Erwin, 9–10
parents: opinion of schools, 24; politici-
    zation of, 13, 139–40; response to new
    math, 1, 123; and textbooks, 2, 105
Piaget, Jean, 88–90
Pike, Nicholas, 59–61
Pólya, George, 61, 70
Price, G. Baley, 42, 73, 93, 118
Progressive Education Association, 29,
    60–61. *See also under* education
psychology. *See under* arithmetic
publishers, textbook, 48, 77, 104, 109–12,
    144. *See also under* School Mathemat-
    ics Study Group

Reagan, Ronald, 25, 139
Rickover, H. G., 26–27
Rosenbloom, Paul: on applied mathemat-
    ics, 67–68; on Cold War education,
    116–17; on elementary students, 89,
    117; on elementary teachers, 105; on
    evaluation of curriculum, 126; as sup-
    porter of SMSG, 41, 99–100, 109

Sarason, Seymour, 127–28
School Mathematics Study Group (SMSG):
    and commercial publishers, 3, 102–4,
    108; difficulty of materials, 106–7; as
    egalitarian, 92–93; and elementary ed-
    ucation, 80–82, 97, 105–8; ending of,
    123; influence of, 2, 45–46, 67; and in-
    telligent citizenship, 3–4, 14, 114; nov-
    elty of, 14–15, 66–67; origins of, 41–42,
    48–49; organization of, 42–44; and
    other reform groups, 15–16, 93–94; ped-
    agogical innovation of, 16; and profes-
    sional mathematicians, 2–3, 15, 46–49;
    psychologists, work with, 90–91; and
    secondary education, 97–99; and state
    boards of education, 100–102, 108–9;
    and student types, 3, 77, 92–93; testing
    centers, of, 99–101. See also Begle, Ed-
    ward; new math

schools: culture in, 127–28; middle, or ju-
    nior high schools, 53–54; primary, or
    elementary schools, 76–77, 82, 105–12,
    123; secondary, or high schools, 11, 97–
    105. See also education
Society for Industrial and Applied Mathe-
    matics, 67–69, 73. See also mathematics
Sputnik, 17–18, 23, 25–27, 34, 39–41
Stone, Marshall H.: on computation skills,
    82; Richard Courant's views of, 67; on
    elementary curriculum, 80; on "mod-
    ern" mathematics, 52–53, 59; on Morris
    Kline's views, 72; psychology, views on,
    87; as supporter of SMSG, 41

textbooks. See publishers, textbook
Truman, Harry, 36
Tucker, Albert, 49, 52

University of Illinois Committee on School
    Mathematics (UICSM), 2, 15–16, 100

Waterman, Alan, 37–39
Wilson, Sloan, 33, 35
Wittgenstein, Ludwig, 79–80
Wooton, William, 103–4

Zacharias, Jerrold, 33, 39